Multilingual Education

CW01501973

The *Multilingual Education Yearbook* publishes high-quality empirical research on education in multilingual societies. It publishes research findings that in addition to providing descriptions of language learning, development and use in language contact and multilingual contexts, will shape language education policy and practices in multilingual societies.

The *Multilingual Education Yearbook* is highly relevant to researchers in language and education, language education professionals, and policy makers, covering topics such as:

- The effects of multilingual education and literacy education on the maintenance and development of multilingualism.
- The effects of the introduction of English as a curriculum subject and/or medium of instruction upon multilingual and literacy education.
- The respective role(s) of vernaculars and 'local' languages, national languages and English in education, especially where the languages are of different language families, and scripts are different or languages lack an orthography.
- The role in multilingual education of other major languages such as Arabic, French, Hindi, Mandarin and Spanish.
- The effects of multilingual and/or English language education on school drop out and retention rates.
- The effects of the 'internationalization' of universities worldwide, potential privileging of the English language and of knowledge published in English.
- Bilingual/multilingual acquisition of non-cognate and 'different-script' languages.
- Takeholder attitudes toward notions of multilingualism and related notions of linguistic proficiency, standards, models and varieties.
- Critical evaluations of language policy and its implementation.

More information about this series at http://www.springer.com/series/15827

Anthony A. Essien · Audrey Msimanga
Editors

Multilingual Education Yearbook 2021

Policy and Practice in STEM Multilingual Contexts

 Springer

Editors
Anthony A. Essien 🆔
Wits School of Education
University of the Witwatersrand
Parktown, South Africa

Audrey Msimanga 🆔
Sol Plaatje University
Kimberley, South Africa

University of the Witwatersrand
Parktown, South Africa

ISSN 2522-5421   ISSN 2522-543X (electronic)
Multilingual Education Yearbook
ISBN 978-3-030-72011-7   ISBN 978-3-030-72009-4 (eBook)
https://doi.org/10.1007/978-3-030-72009-4

This Springer imprint is published by the registered company Springer Nature Switzerland AG
The registered company address is: Gewerbestrasse 11, 6330 Cham, Switzerland

Preface

There is no question that a Yearbook focusing on Policy and Practice in STEM Multilingual Contexts is due. With recent advances in global mobility, all STEM subject contexts have become multilingual in nature. In turn, the role of language in STEM access and success has become the subject of universal concern for educators, researchers, and policy makers alike. This Yearbook is intended for the attention of students, educators, teacher educators, researchers, as well as policy makers. The intention in each chapter in this Yearbook is not to provide answers and solutions but to raise critical controversial and politically sensitive questions to challenge the STEM research and education communities together with the Language education fraternity and indeed the education stakeholders in the policy and political domains. The chapters are presented in a manner that allows the authors to build on each other's arguments.

"Challenges and Opportunities from Translingual Research on Multilingual Mathematics Classrooms" sets the scene by looking back and looking forward at the theory, policy, and practice terrain of multilingual mathematics research, with a focus on the critical relationships between monolingually oriented educational policies and the practicalities of the multilingual mathematics classrooms universally. "Appreciating the Layered and Manifest Linguistic Complexity in Mono-Multi--Lingual STEM Classrooms: Challenges and Prospects" picks up the argument to interrogate the layered linguistic complexities of mono- and multilingual teaching and learning contexts in general, and how the STEM-specific layering in particular complicates the deployment of linguistic resources in the classroom. "Approaches that Leverage Home Language in Multilingual Classrooms"–"Practices in STEM Teaching and the Effectiveness of the Language of Instruction: Exploring Policy Implications on Pedagogical Strategies in Tanzania Secondary Schools" draw on evidence from real mathematics and science classrooms in South Africa and Tanzania to illustrate the elusiveness of approaches that leverage home languages teaching ranging from translation strategies to the more nuanced affordances of "translanguaging" approach that allows languages to be treated as resources for learning in early primary mathematics teaching; to the policy–practice tensions arising as both science and mathematics teachers and learners attempt to draw on the linguistic resources of their classrooms in a context of multiple monolingualism set on a falsely assumed multilingual policy environment.

The next four chapters, "Individual Language Planning for Self-Directed Learning in Multilingual Information Technology Classrooms"–"Using Interactive Apps to Support Learning of Elementary Maths in Multilingual Contexts: Implications for Practice and Policy Development in a Digital Age"–"Noticing Multilingual and Non-dominant Students' Strengths for Learning Mathematics and Science"– "Multilingual Students Working with Illustrated Mathematical Word Problems as Social Praxis" look at specific learner/student experiences of learning science and mathematics in the policy environments described so far. Two studies report on the potential of technological interventions for student learning in a diversity of contexts in three continents highlighting some specific local contextual and more universal technological affordances for mathematics and science learning in multi-lingual contexts. The other two studies delve into the questions of multilingual learner resilience that they bring to the classroom space be it *social, cultural, and linguistic experiences which they draw on to learn specific content or the general* strengths and not deficits which they bring to the classrooms and how important it is for policy and practice to identify and recognize such learner strengths.

The final chapters, "Language Policy for Equity in University STEM Education in Postcolonial Contexts: Conceptual Tools for Policy Analysis and Development"– "The Place Where Languages Meet to Argue: A Contribution from an Analysis of the Brazilian National Curriculum"–"Principles for Curriculum Design and Peda-gogy in Multilingual Secondary Mathematics Classrooms" address the intersections between curriculum, curriculum design, language policy, equity issues, and higher education and/or teacher education. In many postcolonial contexts, language remains the key factor in the continually elusive and highly politically charged subject of inequity in STEM education. Two of the studies interrogate the nuances of access and success in STEM in Brazil and South Africa. The Brazilian study observes a policy–curriculum–practice contradiction which works against the espoused interdis-ciplinarity while the South African study observes a persistent policy focus on access to and achievement in dominant STEM knowledge in "English" with recommenda-tions for future policy development. The USA study on the other hand makes recom-mendations for mathematics curriculum development that aligns the conceptual focus and use of problem contexts across each curricular unit, integrates practice-focused and content-focused learning goals in a trajectory while incorporating structures that enable the widest possible participation and access to science for multilingual learners.

Parktown, South Africa   Anthony A. Essien
Kimberley, South Africa   Audrey Msimanga
Parktown, South Africa

Acknowledgments Our appreciation goes to the Faculty of Humanities Research Committee of the University of the Witwatersrand, for providing the funding that enabled the initial proofreading of the chapters in this book. Our sincere gratitude also goes to Senamile Dlamini for serving as the proofreader of all the chapters in this book.

About This Book

This book engages with how policy is deconstructed and reconstructed in practice in contexts of language diversity. It attempts to foreground how challenges and complexities between policy and practice intertwine in the teaching and learning of the STEM subjects in multilingual settings, and how they (policy and practice) impact educational processes, developments, and outcomes.

This book presents high-quality empirical research on education in multilingual societies. Data-based studies, theoretical/position pieces, and comparative studies from different levels of Education (from Early grades to University) in different multilingual contexts around the world included in this book highlight findings and theorisations that will help shape future language education policy and practices in multilingual societies.

The unique feature of this book lies in its combination of not just language issues in the teaching and learning of the STEM subjects, but also how these issues relate to policy and practice in multilingual contexts and how STEM research and practice may inform and shape language policies and their implementation in multilingual contexts.

Contents

Editors and Contributors

About the Editors

Anthony A. Essien is an Associate Professor and the Head of the Mathematics Education Division at the University of the Witwatersrand, South Africa. He is a Series Editor of the book series *Studies on Mathematics Education and Society*. His field of research is in mathematics teacher education in contexts of language diversity. He is also a current member of the International Committee (Board of Trustees) for the International Group for the Psychology of Mathematics Education (IGPME). Anthony also served as an Associate Editor of *Pythagoras*, the academic journal of the Association for Mathematics Education of South Africa, for 11 years. In addition to his background in mathematics education, Anthony also has a background in Philosophy. ORCID ID: https://orcid.org/0000-0002-0040-8773.

Audrey Msimanga is an Associate Professor of Science Education, currently the Head of Education at Sol Plaatje University as well as a Visiting Researcher at the University of the Witwatersrand. Audrey has worked in Biology research and then in Science Education for over 30 years. Audrey's research seeks to understand the role of social interaction in science learning; the potential for classroom talk to mediate learner meaning-making as well as the role of language in science teaching and learning in multilingual classrooms. Audrey is currently an Associate Editor for the Journal for Research in Science Teaching (JRST). ORCID ID: https://orcid.org/0000-0002-7036-1181.

Contributors

Nathália Helena Azevedo University of São Paulo, São Paulo, Brazil

Bongi Bangeni University of Cape Town, Cape Town, South Africa

Laura Caligari Department of Mathematics and Science Education, Stockholm University, Stockholm, Sweden

Maureen A. Callanan University of California, Santa Cruz, CA, USA

Ernesto Daniel Calleros San Diego State University, San Diego, CA, USA

Opanga David University of Rwanda-College of Education, African Centre of Excellence for Innovative Teaching and Learning Mathematics and Science, Kayonza, Rwanda

Lanaya J. Davitt School of Psychology, University of Nottingham, Nottingham, UK

Renata de Paula Orofino Federal University of ABC, Santo André, Brazil

Anthony A. Essien Wits School of Education, University of the Witwatersrand, Johannesburg, South Africa

Julianne Foxworthy Gonzalez University of California, Santa Cruz, CA, USA

Anthea Gulliford School of Psychology, University of Nottingham, Nottingham, UK

Salvador Huitzilopochtli University of California, Santa Cruz, CA, USA

Evalisa Katabua School of Education, University of the Witwatersrand, Johannesburg, South Africa

Kate le Roux University of Cape Town, Cape Town, South Africa

Sam R. McHugh University of California, Santa Cruz, CA, USA

Carolyn McKinney University of Cape Town, Cape Town, South Africa

Evariste Minani Department of Mathematics, Science and Physical Education, University of Rwanda-College of Education, Kayonza, Rwanda

Judit N. Moschkovich University of California, Santa Cruz, CA, USA

Audrey Msimanga School of Education, Sol Plaatje University, Kimberley, South Africa;
School of Education, University of the Witwatersrand, Johannesburg, South Africa

Eva Norén Department of Mathematics and Science Education, Stockholm University, Stockholm, Sweden

Théophile Nsengimana Department of Mathematics, Science and Physical Education, University of Rwanda-College of Education, Kayonza, Rwanda

Jako Olivier Research Unit Self-Directed Learning, Faculty of Education, North-West University, Mahikeng, South Africa

Laura A. Outhwaite School of Psychology, University of Nottingham, Nottingham, UK;
Institute of Education, University College London, London, UK

Kevin Pelaez San Diego State University, San Diego, CA, USA

Nicola J. Pitchford School of Psychology, University of Nottingham, Nottingham, UK

Núria Planas Universitat Autònoma de Barcelona, Cerdanyola del Vallès, Spain

Manono Poo Wits School of Education, University of the Witwatersrand, Johannesburg, South Africa

Ingrid Sapire Wits School of Education, University of the Witwatersrand, Johannesburg, South Africa

Daniela Lopes Scarpa University of São Paulo, São Paulo, Brazil

Jabulani Sibanda School of Education, Sol Plaatje University, Kimberley, South Africa

Alphonse Uworwabayeho Department of Early Childhood and Primary Education, University of Rwanda-College of Education, Kayonza, Rwanda

Paola Valero Department of Mathematics and Science Education, Stockholm University, Stockholm, Sweden

Hamsa Venkat Wits School of Education, University of the Witwatersrand, Johannesburg, South Africa

Nsengimana Venuste Department of Mathematics, Science and Physical Education, University of Rwanda-College of Education, Kayonza, Rwanda

William Zahner San Diego State University, San Diego, CA, USA

List of Figures

List of Tables

Challenges and Opportunities from Translingual Research on Multilingual Mathematics Classrooms

Núria Planas

Abstract This chapter provides a commentary on multilingual mathematics class-room research over the last two decades in order to look toward the next decades with a focus on issues of policy and practice. I select some of the theoretical nuances and concerns that have shaped the domain with respect to the critical relationships between: (1) monolingually oriented educational policies and progress in multilingual mathematics classroom research; and between (2) this progress and its implications for mathematics teacher education policies and pedagogies. To this end, I undertake a threefold interpretation of progress in the research domain. I argue that three meta-theoretical concerns have challenged, not without frictions and back-and-forth fluctuations, monolingually oriented policies, practices, and theories. I start with domain research grounded on language as tool of communication and on codeswitching as encoder of accurate meaning in multilingual mathematics teaching and learning. I follow with research that interrogates the ideal of meaning accuracy, and then end with the most recent line of translingual domain research with implications for the broader field and the work of teacher educators and researchers.

Keywords Multilingual mathematics classroom research · Monolingual policies · Multilingual learners · Translanguaging practices · Translingual position

1 Monolingual Policies and Practices, Multilingual Learners

Research in multilingual mathematics classrooms has become more and more common in the field of mathematics education, with some of the studies bringing up claims of possible generalization to a diversity of educational contexts and content areas. This chapter provides a meta-theoretical commentary on research and guiding ideas specifically created in the field of mathematics education over the last two

N. Planas (✉)
Universitat Autònoma de Barcelona, Cerdanyola del Vallès, Spain
e-mail: Nuria.Planas@uab.cat

decades in order to look toward the next decades with a focus on issues of policy and practice that can inform work in other educational fields. I do not review the research literature in any systematic sense, but rather select some meta-theoretical concerns and nuances that have shaped the domain concerning the critical relationships between: (1) monolingually oriented educational policies and progress in multilingual mathematics classroom research; and between (2) this progress and its implications for mathematics teacher education policies and pedagogies.

Three points are particularly influential and have consequences for the interpretation of the two abovementioned critical relationships. First, I have carried out most of my research and developmental work in multilingual mathematics classrooms of Catalonia. In this autonomous region in Spain, Europe, Catalan is the language of instruction since 1985. Law allows families the choice of education in Spanish, the other official language, although this rarely happens due to the high status and generalized use of Catalan. This is not the case, for example, in France, where the language policy categorizes Catalan as regional and includes it in the not very popular model of bilingual education available for regional languages. Second, I am trilingual myself, or a double second-language learner. I was brought up speaking Catalan, found Spanish out of the home at school in the late 1970s, and then found English in the process of becoming a researcher, while also for some years being a secondary school teacher who taught mathematics to learners of diverse cultural groups in the 1990s. Third, I have learned immensely from work with colleagues on the discussion of language-in-education policies and multilingual mathematics classroom practices in Catalonia (e.g. Gorgorió & Planas, 2001), Arizona (e.g. Planas & Civil, 2013), South Africa (e.g. Planas & Phakeng-Setati, 2014), and Greece (e.g. Chronaki & Planas, 2018). This network continues to stimulate my thinking and research, not without acknowledging the singularities and sociopolitical backgrounds across contexts. The parallelisms between the power exercised by state and colonizing languages in public domains including education (by means of the privileging of one language) are many. Nonetheless, the landscapes that emerge from the colonial periods of exclusion of the African languages in South Africa are very different from those that emerge from the nationalist state projects throughout Europe and the United States and the policies of exclusion of non-state languages spoken by the majority of people in certain regions.

In the following sections, I start by arguing that "one classroom, one language, one mathematics" policies, pedagogies, and ideologies have complicated the research on multilingual mathematics teaching, learning, and assessment, by making it more difficult to notice the epistemic function of languages other than the language of instruction and of mathematical cultures other than school mathematics. The "one classroom, one language, one mathematics" tradition actually forms a significant part of the foundation for most classroom research in the field of mathematics education. A generation of scholars (myself included) have walked a long journey to view languages, language use, and curricular content in the multilingual mathematics classroom beyond the countable alternation of separate languages. Still today, state policies, pedagogies, and ideologies of monolingual normativity narrow the lenses through which we think about mathematics teaching and learning, and what can comprise mathematically relevant language use in the distinct research contexts.

Despite this, increased conceptual refinements of language as social, support the progressive uncovering of translanguaging practices across languages and cultures in school mathematics teaching and learning. These practices illustrate the varied and creative possibilities of multilingual language use that put communication and participation at the center and eventually challenge or resist norms of linguistic accuracy. In this chapter, I use translanguaging to refer to situations of practice explained as "the deployment of a speaker's full linguistic repertoire without regard for watchful adherence to the socially and politically defined boundaries of named (and usually national and state) languages" (Otheguy et al., 2015, p. 283). At the ideological-theoretical level, I adopt the term translingual to describe the position that captures and defends the language practices of all those who use the linguistic, cultural, and social resources at their disposal to produce and investigate mathematics teaching and learning. Such a position has implications for policies and pedagogies of mathematics teacher education, but also for methodological practices in the broader research field and for research and practice in related educational fields.

2 Monolingually Oriented Educational Policies

In this section, I begin to address the critical relationship between monolingually oriented educational policies and progress in research on multilingual mathematics classrooms, a relationship that can also be read as the story of progress of this research in spite of or because of these policies. In the research field of mathematics education, and compared to what happens with the tradition of researching the role and use of theories (e.g. Planas & Schütte, 2018), there is not a strong tradition with regard to the role and use of policies. It is often explained that the theories we choose either widen or narrow what we do and see, and how we think. A similar relational perspective can be applied to the policies surrounding the classroom sites in which we plan, design, and develop our studies. As occurs with theories, progress in the research of specific policies and in the understanding of the role these policies play in the research processes mediates the construction of the field directions. Multilingual mathematics classroom research must hence be analyzed with regard to the theories undertaken and the policies that shape education.

The study of language policies as a distinct field can be traced back to the early study of the struggle with the state about whose knowledge, experiences, and ways of using language are legitimate in Fishman et al. (1968), and in Ruiz (1984). These pioneering works support the link of language policies with social processes that guide and regulate, in school education, how teachers mediate classrooms and instruction with multilingual learners, and how researchers decide what can be researched that is realistic (and which is more likely to receive research funding and institutional acknowledgment). Accordingly, policies that influence practice influence the spaces of research on that practice as well. Language policies and monolingual ideologies enter pedagogy and research in multilingual mathematics classrooms in the form of language choice, but also through the tacit establishment of norms such as talking, writing, or reading one language at a time, which ultimately narrow the effective use of all the linguistic, cultural, and social resources at the disposal of practitioners and researchers.

Monolingually oriented educational policies and research traditions that, for example, see multilingual speaking and writing as abnormal, specifically influence multilingual mathematics classroom research, its performance, and its communication. In international journal articles that mention the original languages involved in lesson data, for example, sentences about space restrictions on not presenting them or on not doubling the length of the transcripts are typically accepted, with the resulting monolingual representation of the classroom sites. Even more decisively, policies and ideologies enter the domain through the basic questions we are able to ask about language in the multilingual classroom, the claims and data we are able to hold and collect, the conclusions we are able to draw from those data, and the directions we are able to maintain and foresee.

The language-in-education policies and the studies in collaboration in my major research context over the last two decades co-illustrate the general case of monolingual orientations entering and challenging multilingual mathematics classroom research in the form of the choices as to what to ask, claim, and conclude. As in other world regions, the educational policies in Catalonia fabricate classroom multilingualism in terms of separate languages and linguistic differences among groups of speakers and learners. In the organizational attempt to reduce linguistic differences before entering the regular classroom, there are Catalan language support classes or "special classes" in the schools for learners who are new arrivals (for details, see Newman et al., 2013). Vallcorba (2010), the past Director of the "Plan for Language and Social Cohesion" in education, recalls the language policy principles that declare Catalan as the language of instruction at all levels of education, and as the normal vehicle of expression in internal school activities and in those of external projection.

Regardless of how teachers put these policy principles into practice, we find the representation of the regular classroom as a place where the language of instruction is the "normal vehicle of expression," which resonates with the representation of language as a neutral tool of communication. The fact that the school is the only place of use of Catalan for most learners of immigrant families in areas of poverty is disregarded, and the possibility of multilingual translanguaging practices for learning and teaching subject content. The local policy actually portrays the language support classes as "the" resource in spite of their problematic functioning at several practical levels. More often than not, these classes are: converted into school spaces for keeping mainstream learners labeled as disruptive out of the regular classroom; guided by pedagogies of curricular remedial arrangements and simplified language across school subjects; and conducted by teachers who are themselves new arrivals to the school. All these remind us of the lack of professional preparation on contents of language for subject-specific teaching and learning in the region, which is a feature more widely documented across regions (Essien et al., 2016), whose consequences affect education of different content areas. I will return to this topic on teacher education later in the chapter.

3 Progress in Multilingual Mathematics Classroom Research

Perhaps, because of my first-hand experiences as a second-language learner and a secondary school mathematics teacher, my earliest studies aimed to explore multi-lingual practices in the participation of learners from minority linguistic and cultural groups in mathematical lessons of the regular and the support classrooms (Gorgorió & Planas, 2001). Despite the local policy oriented to make learners monolingual in school work, I knew that these learners' languages functioned as resources in classroom peer work (not so much in the whole group), similarly to what seemed to happen in Arizona (Civil, 2007), South Africa (Setati & Adler, 2000), and Greece (Chronaki, 2009). The representational lenses at that time were therefore not too narrow, but still not sufficiently wide to ask questions and collect data about multi-lingual practices other than translation or codeswitching. While the sociocultural stance for school teaching and learning and for multilingualism explained the medi-ational role of switching languages, the ideals of mathematical and linguistic accuracy remained present. More discussion and lesson observation would be necessary to see, legitimate, and analyze the complex sociolinguistic interactions and translanguaging practices in the classroom work.

Below, I present three meta-theoretical concerns in the domain research that have challenged, not without frictions and back-and-forth fluctuations, monolingually oriented policies, practices, and theories. I draw on instances of classroom data and reflections from three studies to develop a threefold interpretation of progress in the domain. The presentation of this interpretation is, in turn, twofold because, for each concern or step, I first introduce evidence from my own empirical research and then relate it to evidence from other studies in the domain literature. Even though I model these concerns separately, using different studies in time, they are dependent on each other and on a diversity of theoretical refinements of language as social (Planas, 2018; Planas et al., 2018). I start with research grounded on the ideal of language as a tool of communication that legitimated the efforts to study the prac-tice (shortened in the literature as codeswitching) of using two or more languages to encode accurate meaning in classroom mathematical conversations. I then follow with domain research that epitomizes the political dimension of language as social through the critique of the ideals of communicational tool and accurate meaning, and of universal school mathematics. The third meta-theoretical concern is domain research that gives visibility to classroom practices of translanguaging that acts as a catalyst against the presumed highest epistemic value of the "one classroom, one language, one mathematics" ideology.

In the regular classrooms of the three studies I use to show my arguments, all the learners with stories of immigration shared at least one common language with teachers and peers. This composition was a result of policies of segregation with Latin American migrants mostly living in some districts of Barcelona, the main city in my region. The planning of newer research projects was shaped by these policies, and by the illusion of classroom sites of simplified methodological and translation work to make sense of the languages and cultures of the learners who did not belong to the majority group. Still today, the local funding policies contemplate the category of technical assistant for specific languages (Amazigh, Bangla, Urdu…), to be contracted for relatively short amounts of time which do not allow them to become interpreters or researchers with whom to discuss meaning in data. Given my capacity for "self-translating" in Catalan and Spanish, the option of work with bilingual sites was a relief at that time. I was somehow limited by a language ideology centered on the naturalization of some languages over others, and on the importance of attending to the details of language separation and "correct" linguistic form, that did not allow me to see and recognize fluid moves across multiple languages in the classroom interactions of my data.

The field of critical linguistics has thoroughly examined and uncovered the (un)articulated language ideologies (of the particular research communities, of researchers, of the school, of teachers, learners, and families) that constrain the vision of the "multi" and discriminate non-standardized languages, cultures, forms of knowledge, and practices (Blommaert, 1999). This notion of language ideologies is a key to understand the powerful role of language in making and changing the world, specifically the worlds inside and outside schools and classroom research, through the naturalization of common-sense ideas such as the need for a hierarchy of bounded, named languages traversed by norms of linguistic purism and other notions of language correctness. Whether or not aware, mathematics education researchers, mathematics teachers, and mathematics teacher educators are not immune to these conflicting, received language ideologies of construction, recognition, and impo-sition of an official culture and language norms. These are ideologies and norms conducive to the enforcement of a homogenous culture of school mathematics and of schooling, and to the devaluation of all other cultures and languages and of their speakers.

3.1   First Concern: Asking Questions About Codeswitching

Planas and Setati (2009) document multilingual spontaneous practices of learners with minimal pedagogical intervention from teachers. In that study, the collection and analysis of classroom data were aimed at investigating "how much" Catalan and Spanish each Latin American migrant learner had spoken during five lessons, and then at "counting" the shifts between Catalan [C] and Spanish [S]. There were two research questions: (1) Do Spanish-dominant bilingual students in Catalan classrooms switch languages during mathematical activity? (2) If so, what are some of the factors that seem to promote the language switching with a group of these students in the context of specific lessons? An assumption framing these questions was that the contrast between mathematical participation in small and whole groups was a consequence of the language mostly spoken in the interaction. We corroborated this assumption and concluded that the learners switched to the home language as soon as the conceptual demands in peer mathematical talk increased. At that time, classroom work in which learners used their languages for mathematical communication in creative, hybrid ways was not studied. The decision of examining language shifts rather echoed ideologies of linguistic purism and language separation. This is a literal piece of the data in peer work (p. 43):

Máximo:	[C] Hem de decidir les fletxes que dibuixem i ja està./We need to decide the arrows that we draw and that's all
Eliseo:	[C] Primer pensem les fletxes, després les dibuixem i després en parlem./First we think about the arrows, then we draw them and then we talk about it
Máximo:	[S] Esta idea de las flechas no es fácil. Tenemos que imaginar los diferentes movimientos que existen dentro del tornado./This idea of the arrows is not easy. We have to imagine the different movements that exist within the tornado
Eliseo:	[S] Una flecha tiene que ser una línea recta para que el tornado baje. Tenemos la *t* para la translación./An arrow needs to be a straight line for the tornado to go down. We have the *t* for the translation

The contents under discussion in the excerpt above are part of the unit called "Our dynamic planet," which included a variety of paper-and-pencil mathematical activities that encouraged learners to pose questions and solve problems about Euclidean geometrical transformations in real contexts. In the lesson of the excerpt, the teacher wanted the learners to think about and graphically represent on the plane the composition of spatial transformations such as translations, rotations, homotheties, and symmetries. The central task in this lesson was "How can you mathematically represent a tornado?", and learners started drawing arrows to represent linear motion. In all the lessons, there was an initial open-ended question presenting the task that had more than one answer.

Sociocultural studies that have approached the multilingual mathematics classroom from a communicational perspective well document the attention to characteristics of alternating or shifting languages. Moschkovich (2002) with Latino bilingual learners in California, and Setati and Adler (2000) with multilingual learners in South African townships, characterized codeswitching as a strategy for convergence toward the academic register of school mathematics in the English language of instruction. Shifting languages was therefore an action with meaning in the interactional processes and not just a simple expression of choice. This field-based characterization of codeswitching helped the domain to understand some of the mathematically relevant functions and dynamic forms of language common to mathematics lessons in research contexts with monolingually oriented policies and pedagogies. In this way, it could be claimed that codeswitching is a "natural" unproblematic consequence of what multilingual learners do with language in the mathematics classroom, even when the educational policies and the classroom norms are differently oriented and refrain learners from flexibly using their languages to express and develop their thinking. It was also claimed that multilingual learners within the mathematics classroom behave as people who speak more than one language generally do, and that codeswitching is not necessarily a symptom of lexical gaps, linguistic difficulties, or deficient language abilities of specific groups of learners.

The location of switching languages on the same communicational continuum as other language practices in multilingual talk reinforced the moves away from remedial views of learners' codeswitching. Multilingual mathematical interaction was now closer to being understood in relation to the multiple linguistic forms available in language for the development of the learner's ability to engage with the content. For this to happen, nonetheless, the interrogation of the ideal of codeswitching as an encoder of both communication and accurate meaning would be necessary. Domain researchers were leaving behind the vision of the languages of multilingual learners as sources of difficulties and were starting to see them as sources of mathematical meaning.

3.2 Second Concern: Asking Questions About Languaging

Planas (2014) documents classroom codeswitching viewed as related to the generation of mathematical learning opportunities in peer work, and hence conceptualized as learning resources and sources of mathematical meaning rather than indicators of linguistic difficulties or lexical gaps. The initial research question of that study centered on the potential of switching languages for classroom mathematics learning and meaning making. In a period in which I had not yet started reading about translanguaging and translingual research, an unexpected finding was very revealing. I could see Latin American learners who codeswitched to talk about the language ("the Spanishes and Catalans") they were producing and about their ways of using it during mathematical work. It seemed as if they were acting on language or languaging by talking about their linguistic innovations and those of the others in the small group. Some of these languaging processes interacted with talk in ways that seemed to unravel mathematical meaning and understanding; thus, they could be read as an expression of language as a resource for mathematics learning. An example reproduced in that paper shows learners who language (act on language) in Catalan (italics) and Spanish (non-italics) to invent and use two pair names without genuine meaning in the school mathematics for numerical consecutiveness. This is a literal piece of part of the transcript for that example (p. 61):

Anna:	*Puc donar exemples, com 3 + 5, 3.5 + 4.5... És sumar 1.* [*I can give examples, like 3 + 5, 3.5 + 4.5... It's adding 1*]
Juan:	*Sí, sumar 1, però els,* los números tienen que ser *no continuats,* como 3, 4, 5, 6... [*Yes, adding 1, but the,* the numbers need to be non-continued, like 3, 4, 5, 6...]
Carmen:	*Continuats?* Es *consecutius!* [{smiles} *Continued? It's consecutive!*]
Juan:	*Conse... Mira, aixi ho entendrem millor.* 3.5 y 4.5 son *continuats.* [*Conse... Look, this way we will understand it better.* 3.5 and 4.5 are *continued*]
Anna:	*Són consecutius per la resta, és 1.* [*They are consecutive because of the difference, it's 1*]

(continued)

(continued)

| Juan: | *Però són decimals... son continuats*, no es que se siguen. [*But they are decimals... they are continued, they don't follow each other*] |
| Carmen: | *Sí, han de ser consecutius i que...* se sigan. [*Yes, they need to be consecutive and...* follow each other] |

An important conclusion was that the discussion around the created terms enabled learners to interact in fluent ways and prompted mathematical reasoning on the classroom task. Learners could rather have focused on abnormal language use and put concerns of linguistic or meaning accuracy above the discussion, negotiation, and naming of a concept of numerical consecutiveness imagined for rational numbers. Instead, they engaged in creative mathematical and linguistic processes. It is far from easy for this to happen and to be seen in the context of naturalized norms that enforce the adequacy of specific school mathematics in the language of instruction, which tend to limit classroom talk and mathematical curiosity. In my role as researcher, I was now ready to focus on mathematics learning beyond evidence of adjustment to prescriptive school mathematics and normative talk. The resulting conclusions about the pedagogic realization of multilingual languaging are common to a stream of sociocultural studies under the approach of language as a resource that is closer to seeing processes of mathematics learning beyond linguistic and meaning accuracy and prescriptive curricula. Martínez (2018) with bilingual participants in language immersion classes in Colombia and in the United States, and Phakeng et al. (2018) with data from South African, Indian, and Catalonian trilingual classrooms, also exemplify languaging events with language inventions and innovative talk about language and some of the positive effects on mathematical work and learning.

The approach in Planas (2014) took a contrastive stance to the approach in Planas and Setati (2009) in that it included the explicit vision of codeswitching as a resource in processes of talk about language and mathematics, and the possibility of unpredictable mathematical meaning construction and language use. Nonetheless, it remained limited by monolingual orientations in a number of subtle ways, such as the choice of how to transcribe mathematical talk. The learners in the data did not speak the mathematical symbols for numerals as I made them appear in the transcripts. I had not considered how learners said *3 + 5, 3.5 + 4.5, 3, 4, 5, 6* or *3.5 and 4.5* in peer work as if that was not relevant data, or the embedded processes of making mathematical meaning could be guessed. The use of italic and non-italic fonts for numerals, as if the meaning encoded belonged to Catalan or Spanish, was theoretically and politically sensitive, since it reflected the representation of a neutral mathematical language with autonomous existence. From the mathematical meaning perspective, it is however very different to say "three and half and three plus one and half," or "three point five and four point five," or even "three with five and four with five," all of which are possibilities coming out of *3.5 and 4.5*. The omission of how learners talked symbols in peer work had undermined the communication of processes of mathematical meaning making and their mathematical ability to discuss numerals and properties like being consecutive in combination with number subsets.

3.3 Third Concern: Asking Questions About Translanguaging

Today, the conclusion that translation and codeswitching are just some of the many multilingual language forms available along the communicational continuum supports domain researchers in the ability to see mathematical, linguistic, and semiotic creations of significance in the interrogation of linguistic and meaning accuracy and universal school mathematics. This includes the coinage of the "trans" terminology in the research domain that functions to mean the conceptualization of the language of (school) mathematics, and not only language or the language of learners or the language of teachers, as diverse and fluid. Accordingly, practices of translanguaging refer to the creative use of language for mathematics teaching and learning as classroom participants make sense of their worlds and identities in relation to the languages and cultures of mathematics of the others. A typical situation of translanguaging is when bilingual migrant learners retain their home languages for peer work discussions, and produce linguistic forms that mix the academic language of instruction and the everyday home languages to talk in ways that create broader spaces for interrogation and understanding of mathematical meaning.

The research questions in Planas and Chronaki (2021) directly relate translanguaging to multilingual hybrid spaces that challenge restrictive views of school mathematics and of linguistic behavior. That study shows how translanguaging in mathematical talk displays linguistic creations through combined forms of everyday and academic Catalan and Spanish to challenge the institutionalization and implementation of naturalized mathematical and non-mathematical meanings. Just as codeswitching had appeared to be a common practice of the multilingual learners in the mathematics classroom rather than an aspiration, linguistic translanguaging also appeared to be an existing condition of mathematical conversations among multilingual peers. In an example, the struggles for meaning around "baixar" [going down] and "saltar" [jumping] became part of the mathematical talk to solve the particularization of a Fibonnaci-type problem by counting the possibilities of climbing a staircase through combinations of one and two step-sizes that add up exactly to ten. This problem, together with another Fibonacci-type problem on seating arrangements with rows of desks in a classroom, were posed to work on divisibility properties and reflect on the mathematical fact that every positive integer can be expressed as a sum of distinct Fibonacci numbers. This is a literal piece of the transcript, now with the linguistic expressions for numerals (p. 159):

Maria:	Sempre es baixa, no t'estàs parat	You are always going down, you don't stand still
Leo:	Pero a veces no bajas, saltas. Y a veces solo bajas	But sometimes you don't go down, you jump. And sometimes you only go down
Maria:	Ada, tu ho tens clar?	Ada, does this make sense to you?

(continued)

(continued)

Ada:	*Sí, baixar*	*Yes, going down*
Ton:	*Et deixes combinacions d'uns i dosos*	*You're missing combinations of ones and twos*
Leo:	He empezado pero hay mucho que bajar y saltar. Al menos treinta. Si la escala fuera más corta…	I began but there is too much to go down and to jump. At least thirty. If the staircase was shorter…
Ton:	*Umm… Si fos tres, seria: u, u, u; dos, u; u, dos… i dos, dos impossible. Ara ve quatre*	*Umm… If it was three, it would be: one, one, one; two, one; one, two… and two, two impossible. Now it's four*

At this point of the progress in multilingual mathematics classroom research, several issues arise. Compared to the spontaneity of learners' translanguaging, translingual pedagogies and translingual research methodologies are complex in contexts of one culture of mathematics and one language of instruction. Even though we can be attentive and sensitive to the politics of our choices in research, we can involuntarily continue to prescribe linguistic differentiation and mathematical universalism, and hence limit the representation of some of the learners' abilities in the data and findings. In Planas and Chronaki (2021), for example, the original lesson transcripts again use italic and non-italic fonts to distinguish Catalan and Spanish; a choice that resonates with a monolingual view of bilingualism and a tacit allusion to language separation. While this choice can be an object of critique, the analytical focus on mathematical content and on processes of mathematical meaning positively aligns with a translingual stance in the research domain. We find similar methodological-analytical concerns in the work of Gándara and Randall (2019) with multilingual mathematics learners' translanguaging in the Democratic Republic of the Congo, and in the work of Garza (2017) with Latino mathematics learners who are bilingual in the United States. Both studies address the latent and sometimes productive tension in classroom talk between how much attention to give to mathematics and to language, and specifically how much attention to give in the research process to the politics of representing diverse languages and mathematical cultures through transcripts that unintentionally suggest issues of bright lines between languages.

4 Implications of Translingual Research for Mathematics Teacher Education

I have already explained that the educational policies dictating the representations of multilingualism in my major research context entail problematic strategies of space separation (i.e. regular and special or language support classes), of curricular separation (i.e. remedial simplified subject contents in the support classes), and of teacher separation (i.e. less-experienced teachers assigned to the support classes). Similar arrangements remain problematic in other world contexts like Arizona, South Africa, or Greece. I now reflect upon the possibilities translingual research offers for policies and pedagogies of mathematics teacher education that can make a change in school mathematics education in a direction different to the one traced by monolingually oriented ideologies.

Essien et al. (2016), and Rangnes and Meaney (2020) show the influence of monolingual orientations on mathematics teacher education policies and pedagogies across countries, and suggest research work to deploy pedagogies of "one classroom, one language, one mathematics" also in place in the teacher education institutions. These authors argue for the inclusion of curricular language content for subject-specific teaching and learning in settings of teacher education and professional development. Elsewhere (Planas, 2021), for sites of mathematics teacher professional development, I present the instructional principle of critically distinguishing and choosing or producing instances of lexical elaboration in classroom teacher talk for the overt communication of conceptual meaning within the algebra of equations. Whereas this study does not directly deal with the multiple languages and mathematical cultures of the learners in the classrooms selected for developmental work, it examines the argument for lexical elaboration in articulation with the argument for drawing on fluid language practices in favor of processes of mathematical meaning making. In this chapter, I additionally claim the possibility of envisaging translanguaging itself as a teaching pedagogy or resource that can help student teachers, teachers, and teacher educators to perform their diverse mathematical identities and those of the others in the teacher education modules. Otherwise, if we keep teacher education practices uncritically aligned with monolingual policies and pedagogies, we will prevent (student) teachers from creatively using their languages and mathematical cultures in the thinking of school mathematics teaching, and we will dissuade them from valuing learners' translanguaging.

Just as in the multilingual mathematics classroom, the dominance of monolingually oriented policies and pedagogies does not totally constrain mathematics teacher training. There is room in the teacher education modules for sociolinguistic hybrid spaces of interaction in which translanguaging can take place, and in which more than one language can be chosen to talk and write about more than one culture of mathematics expressed in more than one language. As a mathematics teacher educator in my university, I have been searching for such translanguaging spaces in the curriculum. My research work in multilingual mathematics classrooms has facilitated this search. Linguistic data with school learners and teachers translating, codeswitching, and more generally translanguaging in mathematical interaction have inspired pedagogic work in mathematics teacher education modules. This includes the opening of new learning spaces with the student teachers to experiment with their languages and cultures of mathematics, and the reflection on how school learners also creatively experiment with their languages and cultures of mathematics in the school classroom.

The issue of mixing languages and challenging normative school mathematics is very sensitive in my training context. For some of the future teachers, the material texts with school learners moving between Catalan and Spanish and inventing mathematical meaning through terms like "non-continued" to discuss a possible concept of consecutiveness in the school register of rational numbers, are first experienced as aberrant uses of the language of instruction and of school mathematics. While student teachers are usually favorable to opening space to learners' participation, it is complex to convince them of the pedagogical, epistemic, and political role of translanguaging in the mathematics classroom. For them, the claim that creative processes of mathematical and linguistic meaning are positive for moving mathematical thinking and learning is somehow counterintuitive. It is also difficult for them to notice the risks of undermining some of the learners' mathematical ideas behind the "one classroom, one language, one mathematics" normativity. However, once they start understanding the nature of the relationship between mathematical thinking and translanguaging, they start to be aware that they are in the module as future teachers of mathematics that will teach mathematics, not the language of instruction or the facts of mathematics only, to a diversity of learners.

What is clear from my observations in the university modules is that linguistic hybrid texts of school lesson data offer an excellent basis on which to discuss what it means to be a mathematics teacher in the multilingual classroom. The discussion on these texts helps to interrogate and deconstruct professional identities while paying attention to the struggles between language policies and the communication and learning of mathematics. In a session with preservice teachers in February 2020 in which I proposed work on the complete episode on consecutive numbers, partially reproduced in the previous section, the major goal was to move our thinking from issues of policy and normativity to issues of mathematics learning and school mathematics. One of the student teachers, Lidia, was a vehement defender of prescriptive mathematical meaning:

> *Consecutive numbers mean something very exact and they* [school learners] *need to know. They cannot start inventing, and talking as they want as if that was not a classroom and there was no teacher.* I see what you say, *that they did that to find ways to solve the mathematical problem, and that they succeeded. But this is...* too risky. *They may not be so lucky with the next mathematical problem.* [My translation of my notes in Catalan on Lidia's talk]

Lidia equated the exactness in the school definition of a mathematical concept with the pedagogic conditions of the language processes for making mathematical meaning and explaining the concept. In my reaction, I asked her the meaning of consecutive numbers, or the meaning that she viewed as exact. Her response offered the opportunity to keep working with all the student teachers on the interrogation of static representations of school mathematics and language in mathematical meaning making. She codeswitched to tell us that two numbers are consecutive when "*són enters i se succeeixen uno a uno*" [*are integers and succeed* one by one]. From there, we started the discussion on the differentiation between "successive" and "consecutive" in school numeracy, and put it in contrast to the synonymy of the two pair words in everyday Catalan. Another student teacher, Òscar, interestingly codeswitched to ask:

> *Could we use a longer definition* [of consecutive numbers], *with words like sub...* subsequent *or every next, so that the definition is more an explanation of the mathematical idea?* [My translation of my notes in Catalan on Òscar's talk]

The distinction suggested between knowing a mathematical definition and producing a mathematical explanation functioned to open up interactional spaces for exploring and valuing Lidia's creative use of language as a resource in her explanation of the mathematical concept. More generally, the pedagogic emphasis on the positive opportunities of sharing nuances in mathematical meaning resulted in the interrogation of school mathematical definitions as finite products, compared to mathematical explanations, which limit the learner agency in the thinking and learning process. While offering a critique of the learners in the school episode and taking the role of the definer of mathematical consecutiveness, Lidia had produced an example of translanguaging to initiate the mathematical explanation of the concept against the boundaries of exact or accurate meaning in school mathematics. In discussing the language for numerical consecutiveness not as static definitive text but as fluid in the process of mathematical meaning construction, we see the pedagogic tension between putting limits on language creativity and increasing opportunities of mathematical understanding.

It takes a process of education and of self-interrogation and self-questioning of cultural boundaries for future teachers to see that they are already living in spaces of continuous translanguaging, and that this is potentially good for their preparation as school teachers of mathematics. More time, more agents, and more developmental actions are necessary for teachers and future teachers to make sense of professional identities that challenge language policies and naturalized views of school mathematics. This reflection is also applicable to researchers, and the process that it takes to see the meaning and implications of monolingual and monocultural orientations for research work.

5 The Translingual Position in the Research Field of Mathematics Education

Throughout this chapter, we have seen some of the lessons that multilingual mathematics classroom research can offer to policy, practice (specifically mathematics teacher education), and research in mathematics education. I have drawn together various threads of the argument about the relationships between monolingually oriented educational policies and progress in multilingual mathematics classroom research. Moreover, I have reflected on how these relationships can mediate the emergence of translanguaging pedagogies in mathematics teacher education and, more generally, the adoption of a translingual position that captures and defends the language practices of all those who use the linguistic, cultural, and social resources at their disposal to produce and investigate mathematics teaching and learning. The translingual position encourages the flexible use of languages because it views them as complementary resources that enrich one another and the educational experiences of all participants regardless of their home languages and cultures. Such a position, however, is not easy to undertake. It confronts institutional policies and programs, and instructional materials with a strong monolingual stance across many world contexts in which teachers and teacher educators are differently aware of the pedagogic and epistemic richness of the multilingualism in their classrooms. There are also many challenges with respect to how the translingual position can enter the broader research field of mathematics education and the work of its researchers. In this final section, the focus moves to some of the implications for mathematics education research of interpreting and integrating findings related to the most recent translingual research on multilingual mathematics classrooms. The consideration of effects of the translingual position in terms of policy, practice, and research is completed in this way. Classroom research on language and mathematics has certainly experienced profound shifts from a logic in which languages are distinct from each other, to a logic that considers languages for how they are creatively negotiated and used in mathematical interaction and communication. A growing body of literature accordingly represents the translingual position in multilingual mathematics classroom research. Here, researchers look at how the languages of the learners, of the teachers, and of mathematics cross boundaries to meet, and interrogate translation or codeswitching thought to encode univocal meaning. The broader research field and even part of the specific research domain, however, have not readily adopted this position. More often than not, studies in mathematics education continue to imply work that assumes the fixity of mathematical meaning in processes of translation and of representation of data for conversation with the international scientific community. The issues that our research work needs to address are many in order to develop methodological approaches aligned with the identification, analysis and recognition of translanguaging as common in data and in research practice.

The opportunity for reflexivity generated by translingual research in multilingual mathematics classrooms and mathematics teacher education settings may help field researchers to see the political and ethical role we all have in the material we generate, and in the hybrid spaces we make possible or constrain in our research projects, decisions, and processes. Since the claim of Morgan (2007) that all mathematics classrooms are multilingual in some sense, the world and its classrooms have become more diverse, but illusions of monolingualism and monoculturalism keep moving the research field. The translingual position might facilitate looking across languages and cultures to capture the meanings produced within and by the research process, rather than seeing meaning as static and uncritically tied to assumptions of "one classroom, one language, one mathematics." Such an approach might also facilitate an understanding that even in apparently monolingual research contexts, the ways we use and see language drive data representation and meaning in the production of claims, findings, and reality.

Mathematics education research cannot remain aloof to the theoretical-methodological issues and positions around language, communication, and culture that the practices and consequences of translating data and of assuming unchanged meaning raise. In this regard, a first lesson from translingual research in multilingual mathematics classrooms is that processes of linguistic translation always involve processes of mathematical meaning making and valuing. A second lesson is that working in more than one language, rather than necessarily implying the experiencing of lexical gaps or linguistic difficulties, can result in spaces for language innovation and newer meaning with the potential for increased mathematical interaction and thinking with the "others." The issues at stake are hence beyond mathematics education research in multilingual sites, and have implications for the quality of the data and findings arising from monolingually oriented studies. The politics and ethics of translation are complex and the task of representation in research is fundamentally problematic. It is not trivial and absent of consequences how the translated representation of the language of some learners may cover home cultures of mathematics, how the translated representation of the language of non-English speaking researchers may cover or devaluate ways of thinking, or how the representation of spoken language in written language may be thought of as untranslated and univocal.

All researchers in mathematics education, not only those explicitly working in or with multilingual sites, should develop responsible and ethical awareness of the many nuances that exist in language and in word meaning, and of the many ways in which language and meaning can be used to create spaces of understanding with implications for interaction with other researchers. In this respect, a third lesson from translingual research in multilingual mathematics classrooms is that, even when it seems that the same language is being spoken, the politics and ethics of language require questions about meaning, communication, and understanding. The experience of progress in the research domain makes me think that multilingual mathematics classrooms and mathematics teacher education that is language responsive will provide field researchers, who apparently live and work in monolingual sites, with tools to address the challenges and opportunities arising from language-based mathematics education research. A crucial challenge is the co-construction of progress and reflexivity in contexts framed by the illusion of becoming a learner under monolingual and monocultural conditions. These concerns around capitalizing on translingual research make total sense in other educational fields that deal with teaching and learning in science and science teacher education classrooms.

Acknowledgments Project grants 2017-SGR-101 (Catalan Government); PID2019-104964GB-100 and EIN2019-103213 (Spanish Government).

References

Blommaert, J. (Ed.). (1999). *Language ideological debates*. Berlin: Mouton de Gruyter.
Chronaki, A. (2009). An entry to dialogicality in the maths classroom: Encouraging hybrid learning identities. In M. César & K. Kumpulainen (Eds.), *Social interactions in multicultural settings* (pp. 117–143). Rotterdam, Netherlands: Sense Publishers.
Chronaki, A., & Planas, N. (2018). Language diversity in mathematics education research: A move from language as representation to politics of representation. *ZDM–Mathematics Education, 50*, 1101–1111.

Civil, M. (2007). Building on community knowledge: An avenue to equity in mathematics education. In N. Nassir & P. Cobb (Eds.), *Improving access to mathematics: Diversity and equity in the classroom* (pp. 105–117). New York: Teachers College Press.

Essien, A. A., Chitera, N., & Planas, N. (2016). Language diversity in mathematics teacher education: Challenges across three countries. In R. Barwell et al. (Eds.), *Mathematics education and language diversity. The 21st ICMI Study* (pp. 103–119). New York: Springer.

Fishman, J. A., Ferguson, C. A., & Das Gupta, J. (Eds.). (1968). *Language problems of developing nations*. New York: Wiley.

Gándara, F., & Randall, J. (2019). Assessing mathematics proficiency of multilingual students: The case for translanguaging in the Democratic Republic of the Congo. *Comparative Education Review, 63*(1), 58–78.

Garza, A. (2017). "Negativo por negative me va dar un… POSITIvo": Translanguaging as a vehicle for appropriation of mathematical meanings. In J. Langman & H. Hansen-Thomas (Eds.), *Discourse analytic perspectives on STEM education* (pp. 99–116). Cham, Switzerland: Springer.

Gorgorió, N., & Planas, N. (2001). Mathematics teaching in multilingual classrooms. *Educational Studies in Mathematics, 47,* 7–33.

Martínez, J. M. (2018). Language as resource: Language immersion mathematics teachers' perspectives and practices. In R. Hunter, M. Civil, B. Herbel-Eisenmann, N. Planas, & D. Wagner (Eds.), *Mathematical discourse that breaks barriers and creates space for marginalized learners* (pp. 85–100). Rotterdam, Netherlands: Sense Publishers.

Morgan, C. (2007). Who is not multilingual now? *Educational Studies in Mathematics, 64,* 239–242.

Moschkovich, J. N. (2002). A situated and sociocultural perspective on bilingual mathematics learners. *Mathematical Thinking and Learning, 4*(2 & 3), 189–212.

Newman, M., Patiño-Santos, A., & Trenchs-Parera, M. (2013). Linguistic reception of Latin American students in Catalonia and their responses to educational language policies. *International Journal of Bilingual Education and Bilingualism, 16*(2), 195–209.

Otheguy, R., García, O., & Reid, W. (2015). Clarifying translanguaging and deconstructing names languages: A perspective from linguistics. *Applied Linguistics Review, 6*(3), 281–307.

Phakeng, M. S., Planas, N., Bose, A., & Njurai, E. (2018). Teaching and learning mathematics in trilingual classrooms. In R. Hunter, M. Civil, B. Herbel-Eisenmann, N. Planas, & D. Wagner (Eds.), *Mathematical discourse that breaks barriers and creates space for marginalized learners* (pp. 277–293). Rotterdam, Netherlands: Sense Publishers.

Planas, N. (2014). One speaker, two languages: Learning opportunities in the mathematics classroom. *Educational Studies in Mathematics, 87,* 51–66.

Planas, N. (2018). Language as resource: A key notion for understanding the complexity of mathematics learning. *Educational Studies in Mathematics, 98,* 215–229.

Planas, N. (2021). How specific can language as resource become for the teaching of algebraic concepts? *ZDM–Mathematics Education, 53.* https://doi.org/10.1007/s11858-020-01190-6

Planas, N., & Chronaki, A. (2021). Multilingual mathematics learning from a dialogic-translanguaging perspective. In N. Planas, C. Morgan, & M. Schütte (Eds.), *Classroom research on language and mathematics: Seeing learners and teachers differently* (pp. 151–166). London: Routledge.

Planas, N., & Civil, M. (2013). Language-as-resource and language-as-political: Tensions in the bilingual mathematics classroom. *Mathematics Education Research Journal, 25,* 361–378.

Planas, N., & Phakeng-Setati, M. (2014). On the process of gaining language as a resource in mathematics education. *ZDM–Mathematics Education, 46,* 883–893.

Planas, N., & Schütte, M. (2018). Research frameworks for the study of language in mathematics education. *ZDM–Mathematics Education, 50,* 965–974.

Planas, N., & Setati, M. (2009). Bilingual students using their languages in the learning of mathematics. *Mathematics Education Research Journal, 21*(3), 36–59.

Planas, N., Morgan, C., & Schütte, M. (2018). Mathematics education and language: Lessons and directions from two decades of research. In T. Dreyfus, M. Artigue, D. Potari, S. Prediger, & K.

Ruthven (Eds.), *Developing research in mathematics education: Twenty years of communication, cooperation and collaboration in Europe* (pp. 196–210). London: Routledge.

Rangnes, T. E., & Meaney, T. (2020). Preservice teachers learning from teaching mathematics in multilingual classrooms. In N. Planas, C. Morgan, & M. Schütte (Eds.), *Classroom research in mathematics education: Seeing learners and teachers differently* (pp. 201–218). London: Routledge.

Ruiz, R. (1984). Orientations in language planning. *NABE Journal, 8*(2), 15–34.

Setati, M., & Adler, J. (2000). Between languages and discourses: Language practices in primary multilingual mathematics classroom in South Africa. *Educational Studies in Mathematics, 43,* 243–269.

Vallcorba, J. (2010). *El pla per a la llengua i la cohesió social a l'educació* [Plan for language and social cohesion in education]. Barcelona, Spain: Generalitat de Catalunya.

Núria Planas is a Professor of Mathematics Education at the Autonomous University of Barcelona in Catalonia, Spain, and she was an Honorary Professor of Mathematics Education at the University of South Africa for two consecutive terms. She is currently a member of the Executive Committee of the International Commission on Mathematical Instruction. She is also Editor-in-Chief of the Spanish Journal of Research in Mathematics Education. She has authored numerous journal articles and book chapters, and has coedited book volumes on sociocultural and political issues of classroom research on mathematics, language, and language diversity. Her career started as a secondary school teacher of mathematics in the urban area of Barcelona, where she first became interested in the language of schooling, in pedagogies for multilingual mathematics learning, in institutional obstacles to social justice and equity, and in the double marginalization of working-class learners with stories of immigration.

Appreciating the Layered and Manifest Linguistic Complexity in Mono-Multi-Lingual STEM Classrooms: Challenges and Prospects

Jabulani Sibanda

Abstract This chapter interrogates the layered linguistic complexity in mono- and multilingual contexts generally, and in STEM multilingual contexts specifically. Although the yearbook is about STEM multilingual contexts, understanding the multifarious linguistic challenges within monolingual contexts, with their 'supposed' linguistic homogeneity, buttresses an appreciation of linguistic challenges in multilingual contexts, generally conceived as linguistically diverse. In monolingual contexts, heterogeneity and linguistic complexity are occasioned by social class engendered vocabulary knowledge gap; emergence of language varieties (lingua fracas) deviant from the standard variety; intra-lingual divergence between conversational and academic language; and the oral–literate language dichotomy/continuum.

Multilingual contexts add more languages into the mix, with their nuanced intra- and inter-lingual diversities. Their linguistic and orthographic distance compromise the deployment of diverse linguistic resources in the classroom. STEM subjects add another linguistic layer by their unique disciplinary symbolic language, unique semantics to familiar words, unique syntactic patterns, and unique technical vocabulary.

The chapter problematises the research–policy–practice dissonance that further complexifies instruction within the STEM multilingual contexts. The chapter argues that, notwithstanding the challenges associated with STEM education in multilingual contexts, there are prospects for viewing learners' divergent linguistic repertoires as resources to be capitalised on, and not problems to be shunned and eschewed.

Key words Code-switching · Linguistic alternation · Monolingual · Multilingual · STEM · Translanguaging

J. Sibanda (✉)
School of Education, Sol Plaatje University, Kimberley, South Africa
e-mail: Jabulani.Sibanda@spu.ac.za

© The Author(s), under exclusive license to Springer Nature Switzerland AG 2021
A. A. Essien and A. Msimanga (eds.), *Multilingual Education Yearbook 2021*, Multilingual Education Yearbook,
https://doi.org/10.1007/978-3-030-72009-4_2

1 Introduction

In this chapter, I argue how Science, Technology, Engineering, and Mathematics (STEM) classrooms are linguistically complex spaces, both in what might be considered mono- or multilingual contexts. I also argue that the layered linguistic systems, rendering the classroom linguistically heterogeneous, coalesce into one linguistic system for the individual language user, and represent a resource rather than an impediment to epistemological access. I explore the languaging policy–research–practice interface in multilingual STEM contexts that further complexifies instruction within the STEM multilingual contexts. I further adapt Clarkson and Carter's (2017) framework, meant for generating significant research questions, for the application of inclusive languaging practices in the multilingual STEM classroom, to capitalise on individual learners' linguistic capital.

STEM instruction within the South African context seems to proceed on the assumption that learners are conversant in the Language of Learning and Teaching (LoLT), and that they just need to master STEM disciplinary content. The first sections of the chapter overview the multiplex linguistic networks characterising the STEM multilingual classrooms. An understanding of the linguistic complexity in monolingual contexts heightens an appreciation of the layered complexities in monolingual and multilingual STEM contexts.

2 Linguistic Complexity in Monolingual Contexts

Clarkson and Carter (2017) acknowledge social factors' occasioned linguistic diversity, even where both the teacher and learners share the same language. From a review of several studies, Hurt and Betancourt (2016, p. 4) identify a plethora of environmental factors influencing children's language outcomes, namely; "social and parental support structure, parenting style, maternal speech, nutrition, toxin exposure, exposure to violence, and other prenatal and postnatal stressors", and cite research in twins which shows the greater impact of the environment over genetics in determining language development in low socio-economic environments. Social class and genetics conspire to determine children's language proficiency. As Fernald and Weisleder (2011, p. 2) note, "Claims that early interactions between parents and infants lay the foundation for children's later language and cognitive development are no longer dismissed as scientifically questionable and culturally disrespectful". Monolingual classrooms, therefore, comprise children with differential exposure to rich language, facilitative of concept development and mental connections with things; which approximates the linguistic capital the school draws on. The learners' resultant diverse vocabulary repertoire levels, conversational patterns, and facility to generate language, render the monolingual classroom linguistically heterogeneous.

Cummins' (2008) work distinguishes conversational from academic language, with proficiency in one not necessarily indexing proficiency in the other. Their distinction is in terms of purpose and context of use, occasioned by the use of different lexical, syntactic, and semantic patterns. This places conversational and academic language at two ends of the continuum, with learners coming into the classroom with diverse proficiency levels (at different points of the conversational-academic language continuum) within a language. Such heterogeneity renders the designation 'monolingual' imprecise.

Within-the-language diversity is also occasioned by the mode continuum; a trajectory of language development from informal oral speech to formal academic written language. The density, lexical and syntactic complexity, as well as recourse and non-recourse to prosodic and non-linguistic information of the two modes, make them linguistically diverse. Even within the same mode (spoken or written), there are degrees of formality and structure which require the deployment of diverse linguistic resources, e.g. playground talk versus oral discussion of an experiment. In this example, the diversity in a single medium meshes with the conversational-academic language continuum, making the monolingual classroom a linguistically heterogeneous space.

Diversity in monolingual contexts also manifests through dialects, sociolects, and registers occasioned by the extra-linguistic factors determining language use. Dialects are regional (geographic) or ethnic varieties, whereas sociolects are social varieties determined by socio-economic status, education level, profession, age, ethnicity, and gender, among others. Dialects normally embody unique lexical, syntactic, and phonological subtleties which render them languages within a language. For languages with several dialects, mutual intelligibility between different dialect speakers decreases and even gets lost as one moves from say the first to the last dialect on a continuum. Standardisation, where a dialect is imposed as a standard variety and enjoys prestige, usage, intellectualisation, and codification, is usually politically informed; and in South Africa, it was an apartheid ethno-linguistic project. Register and style, varieties which respond to specific prevailing communicative functions and settings, add to the within-a-language diversity, which further narrows down to the idiolect level (considered later in this chapter), where individuals have their own unique language usage patterns. These monolingual sociological variations render the monolingual classroom, linguistically diverse; a diversity which heightens in multilingual contexts.

3   Linguistic Complexity in Multilingual Contexts

Multilingualism and plurilingualism, distinct and highly contested terms, denote multiplicity of languages within a social context, and the language user's proficiency in multiple languages, respectively (King, 2018). Valid indicators of multilingualism are "...the extent to which there is interaction between linguistic communities, the degree of public acceptance of and support for linguistic diversity, and the ways in which this 'multilingual capital' is part of the political and economic infrastructure, including in the all-important area of education" (King, 2018, p. 8).

In South Africa, the constitutional conferment of official language status to African languages, has further entrenched multilingualism in the classroom. South Africa's indigenous languages belong to two major groups; the Nguni-Tsonga languages (isiNdebele, isiXhosa, isiZulu, siSwati, Xitsonga) and the Sotho-Venda languages (Sesotho, Setswana, Tshivenda). The groupings become four if Xitsonga and Tshivenda are disaggregated to stand independently. The language groupings are an acknowledgement of their linguistic distance. Illustrative is Wet et al.'s (2007, p. 159) observation that "Sotho languages share a system of seven vowels, whereas the Nguni languages have a common five-vowel system". The linguistic distance between and among the indigenous languages impacts even the resultant English variety speakers of the different languages will develop; hence, the designations Sotho English and Nguni English. The linguistic diversity is heightened as each language brings to the mix, several dialects.

Different language groupings dominate specific geographical regions, with Gauteng, the most linguistically heterogeneous, having the two major language groupings represented. That has culminated in the emergence of an argot, Tsotsi-taal. Being relatively young language forms, tsotsitaals are creoles, distinguished mainly by their lexicon to the point of over lexicalisation, where a wide range of words have a single referent. They map onto base language forms and borrow forms and meanings, as well as manipulate the phonological, morphological, and semantic aspects of the base forms to create novel lexical items (Gunnink, 2014). These deviant varieties, which defy lengthy natural language change processes and phonetic prin-ciples, add to South Africa's "multifarious classroom language situations" (Childs, 2016, p. 24).

Sierens and Avermaet (2014, p. 18) posit that "Multilingualism is a motley crew of different, unequally divided competences. Every aspect of language is specifically functional: mastering something in one domain doesn't guarantee success in another domain….". Language is not static, homogeneous, or monolithic, and its acquisi-tion is neither deterministic nor linear. The linguistic complexities characterising multilingual contexts aggravate in multilingual STEM contexts.

4   Linguistic Complexity in STEM Subjects

STEM occasions another linguistic layer by introducing dense technical jargon qualifying as a register or discourse. STEM does not embody universal language independent of linguistic variations. The word 'quadrilateral' is mathematical, and (though of Latin origin) has been incorporated into the English lexicon. It is, there-fore, an English Mathematical term. While two languages (Mathematical language and English) coalesce in the term, one needs to have mastery of the subtleties of English multisyllabic word reading to allow for the word's knowledge at the word recognition level, and one also needs to understand the basics of mathematical shapes and sides to understand the word at the passive or active word knowledge level.

The linguistic complexity heightens when one considers that the disciplines that coalesce into STEM bring in unique technical and symbolic language that renders STEM an amalgamation of four technical languages, over and above the other linguistic diversities already discussed. The discipline-specific STEM language radically shifts the everyday meanings of words, e.g. 'of' taking on new meaning in ¼ of 12. The symbol and graphic (tables, graphs, figures, etc.) density in STEM texts (e.g. pH, $\geq$, $\varnothing$) add another decoding layer as the graphic elements serve communicative not ornamental functions. For Clarkson and Carter (2017, p. 238) "incorporation of many symbols and the truncating of sentences are also elements of the written STEM language quite different to everyday language…". STEM even combines syllabic, logographic (morphosyllabic), and alphabetic writing systems. STEM is a unique language that needs to be mastered.

While it is common knowledge that STEM subjects are a unique language on their own, learning to read the STEM language is neither overtly/systematically taught (but taught simultaneously with the content), nor is the teaching of STEM language reading supported by a body of research. Assuming that, as learners are learning to read in English (for example), they are also learning to read in content areas; is a negation of the distinction between English and STEM language. The common practice in STEM is to teach symbol and graphic literacy as one encounters the symbols or graphs during instruction. STEM instruction in multilingual contexts should be a fusion of languages (multiple languages represented in the classroom and STEM language) and content.

The adage 'Every teacher is a reading teacher' is premised on the twin assumptions of precursors of reading attainment being universal, as well as on cross-linguistic transfer of reading elements across languages (Cummins, 2008). The question to ask is; how well equipped is a STEM teacher to handle the intricacies of vocabulary, fluency, and reading comprehension in STEM teaching? "Mainstream, content area teachers need knowledge and practical ideas about addressing the academic language needs of ELLs because they have the dual responsibility of facilitating ELLs' content learning, while also supporting their ongoing English language development" (de Oliveira, 2016, p. 218). In the STEM context, language (in its multiple and layered manifestations) has to be learnt simultaneously with STEM content, not separately or sequentially. What to foreground and background in this tenuous balancing act needs consideration.

5 Summing the Layered Linguistic Complexity

Clarkson and Carter (2017, p. 240) aptly sum up this multi-pronged linguistic complexity in multilingual STEM classrooms as occurring at:

- "Different 'levels' of language (families of languages, distance between languages)
- Different language contexts (indigenous, multilingual, immigrants)
- Contexts within language (speaking, listening, writing, reading) as well as the immediate context (conversational compared with academic)
- Content realities (cultural, social, political)"

The absence of a shared spoken language outside the LoLT, the growing intersection between and among (coupled with lack of mastery in) the STEM subjects' scientific and technical discourse, and context-specific word meanings, add to the matrix of linguistic challenges in the multilingual contexts. All these linguistic diversities conspire with the other non-linguistic diversities like socio-economic class, to create a mosaic of diversities and confluence that are attractive on the surface but complex to navigate. There is a need for policy and research to inform practice.

6   Policy–Research–Practice Dissonance/Confluence

This section explores the South African policy provisions in relation to the complexities of multilingual classrooms; the contribution of research; and the extant multilingual classroom practices to determine the areas of discordance and areas of consonance.

6.1   Policy

The South African Schools Act's (No. 84 of 1996) devolution of school language policies to school governing bodies (SGBs) did not achieve the desired end of additive bilingualism, as SGBs advocated even earlier introduction of English than under apartheid. With the hegemony of English in learning, commerce, and administration; the school governing bodies are naturally predisposed towards recommending an English-only instructional approach. The constitutional provision in the South African Schools Act (No. 84 of 1996) for learner instruction in their Home Language where 'reasonably practicable' is circumvented by the discretion exercised by the SGBs. Probyn (2017, p. 7) posits that the intuitive assumption that.

> early submersion in English is the most effective way to acquire English… appears to have overridden the paradoxical reality that such policies actually have limited learners' access to the content of the curriculum and have instead blocked them from the desired upward mobility.

This, is despite a voluminous body of counterintuitive research evidence to the contrary.

The South African Constitution Section's 29(2) provision for the education of learners in their home language where 'reasonably practicable' is a veiled acknowledgement of practical constraints that attend the elevation of African languages to official languages. Equity and redress of past imbalances seem to actuate policy provisions and not feasibility concerns. The phrase 'reasonably practicable' cannot be defined with precision, and so English continues to hold exclusive sway in the STEM multilingual classroom on the pretext of any other linguistic innovation being either unreasonable or impracticable.

Stoop (2017, p. 8) notes that "Section 29(2) provides expressly for single-medium institutions…within a range of possibilities….". The right to education, which may best be served by an incorporation of diverse linguistic resources is curtailed when the SGBs, either sideline the African languages or the minority languages and go for a single medium offering.

The English Across the Curriculum (EAC) initiative by the Department of Basic Education is an acknowledgement of the multi-layered linguistic barriers to accessing content. The intervention is meant to develop the twin language arts of listening and speaking, reading and viewing, writing and presenting, as well as language structures and conventions; within content areas that include STEM. Teacher Education institutions have not approached the EAC in a uniform way, with a significant number of institutions known to the author relegating the EAC module to English Education lecturers. This deprives content area lecturers of the knowledge of mediating language and content to their students, and students hardly see the relevance of EAC to their specialisations. The EAC initiative itself is a monolingual intervention to a multilingual and linguistically multi-layered challenge. While the EAC is the policy innovation that comes close to linking language with STEM subjects by virtue of the 'across the curriculum' designation, it is all about entrenching the hegemony of English as both LoLT and a subject and does little to recognise the nuanced STEM linguistic demands.

Robertson and Graven (2020) identify three orientations in language policy and practice debates in multilingual contexts namely; language as problem, language as right, and language as resource. The three orientations can be unpacked as follows:

- Language-as-problem orientation stems from a deficit conception of minority languages and the need for expediting minority language speakers' proficiency in the LoLT. Such assimilationist orientation leads to subtractive bilingualism. In South Africa, the SGBs (and ironically teachers, learners, and parents) by their preference for straight for English practice operate at the language as problem level.
- Language as right seeks to equalise education access to all linguistic groups and engender acceptance and tolerance of previously marginalised languages. In South Africa, the elevation of indigenous languages to official status, and to LoLT status at the Foundation Phase is representative of language-as-right orientation. South Africa's policy positions follow this orientation.

- Language as resource recognises multilingualism and linguistic diversity as desirable and inherently good, and meriting application in the classroom. This chapter advocates adoption of language-as-resource orientation.

None of the extant acts and policies that make pronouncement on language recognise the language continuum or linguistic mosaic pivotal to communication within multilingual classrooms. The acts and policies still reflect the linguistic purism and protectionism notion where the designated 'standard' language should guide classroom discourse. The policy provisions treat languages as separate and bounded systems and do not even hint at the possibility of any language alternation practices in the classrooms to mirror real-life languaging practices.

6.2 *Research*

In as much as the first section of this chapter showed how multilingualism can manifest in a monolingual context, research attests to how monolingual practices have encroached into bi- and multilingual education programmes. The latter has manifested in what has been variously designated "multiple monolingualism" (Sierens & Avermaet, 2014), "two solitudes" (Cummins, 2008), "double monolingualism" (Wedin & Wessman, 2017), and "pluralisation of monolingualism" (Makoni & Pennycook 2007 in Makoe & McKinney 2014, p. 22). Extant mainstream approaches to bilingualism have variously been referred to as 'parallel monolingualism', 'bilingualism with diglossia', 'separate bilingualism', and 'bilingualism through monolingualism'; to show the exclusivity and prescriptive language that must be conformed to, rendering the bilingual "two monolinguals in one body" (García et al., 2011; Makoe & McKinney, 2014, p. 4). King (2018), observes that a bilingual person is not a fusion of two monolinguals in one, where each language retains its separate and independent culture. That is why Probyn (2017) proposes the adoption of a divergent heteroglossic outlook from one that visualises languages as two or more 'inflexible solitudes', to one that recognises a fusion of linguistic forms and repertoires from different languages into one system. Viewed this way, the multiple linguistic resources discussed earlier, characterising both mono- and multilingual contexts, coalesce into one unitary linguistic system rather than multiple and separate language systems. The classroom should dismantle language ideologies and regimes that circumscribe multilingual practices in the classroom and "… homogenise learners and their language practices, reducing complex heteroglossic language use to neat descriptions of full proficiency or lack of proficiency in a named language" (Makoe & McKinney, 2014).

People's languaging defies definitional parameters set by named languages. What Garcia and Otheguy (2020) find unfortunate is how the languaging practices of 'monolingual white elites' are considered the norm. In schools, such manifest power differential "has led to a reductive situation where recognition is only accorded, in a multilingual repertoire, to the use of one, two, or three separate, standardized named language(s)" (p. 18). By valuing and endorsing the standardised varieties of named languages, the school denies the complexity of society's languaging practices rather than capitalise on them; thereby creating dissonance between the school and real-world languaging practices.

In their different ways, plurilingualism and translanguaging challenge the traditional conception of multilingual, monolingual, and monoglossic practices characterising extant language education practices, and seek to leverage learners' linguistic resources. Plurilingual competence is the linguistic repertoire and proficiency in several languages, to varying degrees and for distinct purposes; allowing the language user to deploy the dual and distinct repertoires as and when needed. Plurilingualism occurs through polylanguaging, that is, employing resources associated with diverse languages despite one's limited proficiency in the languages in question.

Translanguaging combines the linguistic, semiotic, and multimodal meaning-making repertoire "as a single inventory of lexical and structural resources, a unitary linguistic system... that they build through social interactions of different types, and that is not compartmentalized into boundaries corresponding to those of the named languages" (Garcia & Otheguy, 2020, pp. 24, 25). The same authors see translanguaging as political and radical, denigrating the legitimated hierarchies of named languages courtesy of racialised, classed, and gendered socio-political categorization; which serve to subjugate minority language communities, compelling them to utilise specific named languages (Garcia & Otheguy, 2020). In translanguaging, the languages known by the language user (at the different levels and dimensions) constitute a single, complete, indivisible linguistic repertoire rather than separate, dual, truncated, structured, and named languages the user comes in and out of as they communicate. They form the language user's idiolect. "Adopting a translanguaging stance and designing translanguaging instruction de-naturalizes the standardized named languages of school. It de-naturalizes, that is, the named languages that have been codified by the nation-state to develop governable subjects" (Garcia & Otheguy, 2020, p. 27). Translanguaging challenges the exclusion of minority language bilinguals' linguistic and cultural capital in the classroom languaging practices occasioned by power hierarchies which compromise and constrain the minority language bilinguals' epistemological understandings and visibility.

Probyn (2017) advocates pedagogical translanguaging, where there is a threefold movement from home language to general language of the First Additional Language (FAL), and to the academic language of the FAL. Within such translanguaging, concepts are deliberately and systematically developed in the Home Language (HL), then transferred to everyday English, and ultimately, to the scientific or technical English. Freeman and Freeman's (2007) Preview-View-Review strategy is an example of deliberate language planning for the multilingual classroom. The preview stage introduces the topic in the HLs (making connections, brainstorming, etc.), the view stage generates details for the topic through the LoLT, and the Review is done in the HL. Such teaching for transfer does not just require proficiency in the HL, the conversational and academic language of the FAL on the teacher's part, but "simultaneously scaffolding a shift across modes from oral to written text production" (p. 14). Such translanguaging, while acknowledging general and academic language as distinct, falls into the concept of named bounded and separate languages the interactants get in and out of.

While plurilingualism would support the strategic scaffolding of one language by another, drawing on unbalanced language repertoires; translanguaging capitalises on learner agency to deploy all communicative resources to transact with texts and with others. One challenge though is that the texts themselves, particularly the print texts, follow the separate named language pattern. Garcia and Otheguy (2020, p. 32) advise the need "to keep the conceptual distinctions between plurilingualism and translanguaging at the forefront as we develop ways of enacting them in practice, even when pedagogies may turn out to look the same".

6.3   Practice

While polylanguaging and translanguaging practices hold the greatest promise in the STEM classroom, research (Clegg & Afitska, 2011; Wedin & Wessman, 2017) attests to code-switching being the most prevalent language alternation practice in the classroom, manifesting as a continuum between propensity towards the base (LoLT) or the embedded form (HL). It is for this reason that this section on language practices predominantly focuses on code-switching.

Although it is a bi- or multilingual practice, in practice code-switching is the momentary switching into alternate language(s) and back to the base form. The teacher engages in a long stretch of monolingual talk in the base form, which s/he punctuates with occasional words and phrases from alternate language(s) or the embedded forms. An example by Clegg and Afitska (2011) is where the teacher talks in the base form, learners conduct group or pair work in the embedded form, and the class holds a post-group discussion activity through the base form. In the majority of cases in South Africa, African languages (which are mostly the embedded forms from Grade 4 upwards) merely provide the brief intervening stretches while English as LoLT is the dominant and base form. Within the STEM context, the technical terms which carry subject content or concepts are, in my experience, given in the LoLT; possibly for lack of equivalents in the embedded forms. Translation, like code-switching, equally represents responsive "… temporary excursions from the monolingual ideal" (Childs, 2016, p. 24). It still falls back into double monolingualism and does not represent the intricacies and creativity of language interactions in the classroom. With time, learners cease attending to instruction in the target language and wait to attend to the easier and translated form.

Although code-switching is less disruptive of language purism ideals (as it recognises the independence of the named languages and their standard forms), stigma still lingers around any language alternation. In my view and experience, sometimes the Home Languages only serve pedestrian non-pedagogical functions like bonding with learners which explains their generous use in class management functions, where the stakes are low. Such classroom language power differentials render African languages "…de facto minority languages in relation to English" (Heugh

2014 in Probyn, 2017, p. 2) despite the numeric dominance of the African language speakers in the South African classrooms (approximately 80% according to Probyn, 2017). The classroom becomes a microcosm of society where "…in Africa, local languages function along horizontal axes, for the purposes of social cohesion and cultural expression; and former colonial languages function along vertical axes for the purposes of the formal economy and politics and are generally learned in school" (Heugh 2014 in Probyn, 2017, p. 2).

Code-switching is normally employed in an ad hoc, spontaneous, unpremeditated, relatively brief, reactive way within a largely monolingual orientation. It is mostly a repair strategy employed when communication fails in the monolingual mode, or for concept clarification or for comprehension check, hence, its prevalence at the introduction than revision of new concepts stages, owing to its concept clarification function as Clegg and Afitska (2011) observe. Lack of pedagogical planning in relation to the employment of code-switching (which can be at the word, phrase, clause, sentence, or beyond sentence level) or code-mixing (which is essentially sentential), culminates in failure to fully capitalise on learners' linguistic resources.

My experience is that code-switching is largely proscribed to the oral component; with reading, writing, and assessment conducted in the base form. That explains teachers' easy-going placatory attitude towards learners' oral expression contrary to their hard uncompromising stance for written expression which should be in accordance with the standard variety of the base form. Sometimes code-switching merely serves a time-saving function where the teacher throws in a word or phrase in the embedded form to avoid lengthy explanation of the same in the base form.

Clegg and Afitska's (2011) distinction between hetero-facilitative and self-facilitative language alternation where the former is actuated by the desire to bring clarity, and the latter by the speaker's limited proficiency (inhibitions) in the base form, is instructive. Both belie a monolingual framework where the embedded form is only imported to solve a difficulty (either the speaker or the hearers') and not as a sound bi- or multi-literacy practice. Sometimes, only the teacher has recourse to the embedded form and, as Wedin and Wessman (2017) observe; mainly to rebuke or on the pretext of clarifying things. Translanguaging or polylanguaging becomes a learner deviant practice done in whispers or in the teacher's absence. 'Deviant' teachers surreptitiously and under cover, smuggle the HL into classroom discourse at the risk of censure for the illicit or transgressive subversion. Such a practice is consistent with language-as-problem orientation rather than language as a significant resource learners take with them to school.

The assessment regimes operate almost exclusively under a monolingual frame. Lopez et al. (2014) observe that "Most content assessments reflect a monolingual or monoglossic or fractional view of language and tend to ignore the complex and discursive practices used by bilingual speakers". Within the monoglossic or fractional perspective, the bilingual is two monolinguals "…with access to two detached language systems that develop in a linear fashion and are assessed separately from one another" (ibid.). The monoglossic assumption is that, despite having facility in the two languages, they can only work in and through one of them at a time. The South African assessment regimes are consistent with such monoglossic expectations which militate against creative deployment of learners' linguistic resources in the classroom. Heteroglossic assessment approaches, however, recognise all languages as part of an integrated system that can be mixed and matched. A brief look at more misconceptions constraining teachers' linguistic behaviours and innovation as they navigate the dicey and exigent language issues in the multilingual classroom follows.

6.4 Constraining Misconceptions

A monolingual English-only approach has been occasioned by the view of the classroom being the sole source of English input for the majority of learners, compelling teachers to plod on with English-only instruction even where language alternation has prospects for greater benefits. Further to that, language purity, verbal hygiene, and fear of negative interference between and among named languages' structured domains, accounts for language education's separation of bilinguals' languages, which explains the 'two monolinguals in one body' concept. Teachers' views of language as a bounded and pure system (Childs, 2016), informs their consternation for assistive language alternation strategies which consequently compromises linguistic and conceptual development. Makoe and McKinney (2014, p. 4) reiterate that "It is the ideology of languages as pure and bounded that underlies the guilt commonly expressed by teachers who do use codeswitching in classrooms where the language of learning and teaching is English, despite English not being the home language of learners".

Language purism is counterproductive "… particularly in urban areas such as in Gauteng Province where there is not a dominant local language, where there are urbanized varieties of African languages spoken that differ from the standardised written forms, and where many children speak hybrid varieties such as 'tsotsitaal' (literally, gangster language)" (Probyn 2015, p. 11). Language purism is assimilationist and, according to Probyn (2017), represents a reproduction of apartheid policies of 'Anglonormativity' at the expense of multilingualism.

Schwarzer and Acosta (2014) identify as a misconception, the view that monolingual teachers are incapacitated to foster multiliteracy on account of not being plurilingual themselves. While the value and expediency of plurilingual teachers cannot be downplayed, the misconception potentially stems from an equally faulty understanding of the role of the teacher as the dispenser of knowledge not an organiser of learning experiences. The latter role allows the teacher to envision prospects for language alternation in the multilingual classroom, and organise for the same.

7 Prospects for Language Use in Multilingual Contexts

Clarkson and Carter (2017) envisage an interplay between some broad theoretical aspects offering a framework for generating significant research questions. These are:

- The structural relation between language and STEM
- The registers and discourses relating to STEM
- The interactions in STEM classrooms
- The different theoretical tools and approaches (p. 240)

In the subsequent sub-headings, I adapt these theoretical constructs, meant for research, to frame STEM instruction in the multilingual classroom.

7.1 The Structural Relation Between Language and STEM

The logic underlying STEM is that of integration, and insisting on a strict monolingual trajectory is anathema to that thinking. If a functional relationship exists among individual disciplines that constitute STEM, a similar structural relation between language(s) and STEM should be acknowledged. As STEM amalgamates its disciplines, its instruction should similarly integrate diverse linguistic resources.

Referring to STEM, Bergsten and Frejd (2019) argue that, there needs to be a balance between; on one hand, ensuring subjects are merged and coherent while retaining their distinct individuality; and on the other hand, ensuring individual disciplines do not just service other disciplines. That structural relationship wholly extends to the multifarious linguistic resources in the multilingual classroom.

The envisaged integration in STEM multilingual contexts should be at the STEM level, at the language(s) level, and at the STEM-Language(s) interface. These integration levels coalesce easily through *multidisciplinary* (themes), *interdisciplinary* (fusing concepts and skills), and … *transdisciplinary* (connected concepts applied in projects and realistic problems) instruction (Bergsten & Frejd, 2019) while translanguaging in classroom interaction. STEM language revolves around problem solving using the scientific method that occasions observations, questioning, experimenting, hypothesising, robust discussion, and collaboration; which heightens the linguistic demand of the STEM classroom.

7.2 The Registers and Discourses Relating to STEM

Word knowledge is indispensable to all learning and classroom communication, and there is a need to systematically determine and delineate, for STEM, the written and oral academic vocabulary needs of learners at particular levels. Sibanda and Baxen

(2016) discuss a principled approach to the determination of the vocabulary needs of learners using a textbook corpus. The corpus can be broadened to include the African language requirements and the oral language corpus. The resultant vocabulary needs can then be explicitly developed at the receptive and productive levels. Vocabulary thresholds should be set for the indispensable vocabulary; which learners should cross to ensure reading to learn. Vocabulary development should neither be imprompt or an afterthought, but deliberately planned and applied to engender contextual rather than general proficiency in the LOLT.

Lefever-Davis and Pearman (2015) envisage the development of strong literacy skills as lying at the heart of promoting STEM learning in a multilingual classroom. The literacy needed in the STEM classroom transcends mere development of requisite domain-specific vocabulary knowledge, to the ability to "interpret and analyze multiple types of texts as well as the ability to express those understandings in creative ways..." (Lefever-Davis & Pearman, 2015, p. 62). This is why relegating English across the curriculum module to the Language lecturers alone risks having the STEM teachers ill-equipped with language-related literacy practices for a STEM classroom. The literacy practices should be reflective of diverse thinking and communication ways in different fields.

Although all languages are capable of communicating any meaning, they need to be adequately intellectualised. It is axiomatic and sobering that, in South Africa, the African languages' modernisation, regularisation, and codification to sufficiently carry out the function of LoLT has not been extensive across all official languages and contexts. This, however, does not preclude them from having a consequential role in STEM teaching and learning. If anything, the proscription of the African languages' roles to non-academic roles is what stifles and delays their intellectualisation and growth. Their grammatical codification should also be informed by their extant and actual classroom usage.

7.3   The Interactions in STEM Classrooms

A range of linguistic and non-linguistic meaning-making resources should be deployed and shared in the multilingual STEM classroom; considering learners' backgrounds and foregrounds within a network of practice (Clarkson & Carter, 2017). Language mediates participation in any classroom context, and in the multilingual context, it becomes prerequisite for inclusion. Exclusive use of the LoLT, language isolation, and language over-regulation may be disempowering and curtail dialogue, engagement, and conceptual development. Focus should be on education of learners equitably and optimally rather than imposition of linguistic norms and practices. Learners should have the autonomy to leverage their linguistic resources to increase comprehension. García et al. (2011, p. 397) rightly observe that "[I]mposing one school standardized language without any flexibility of norms and practices will always mean that those students whose home language practices show the greatest distance from the school norm will always be disadvantaged". Intersecting languages in the classroom is simply "… expanding a multilingual repertoire of different genres, styles, registers and linguistic tools…" (Sierens & Avermaet, 2014, p. 18) which learners already embody.

In everyday life, languages intersect and overlap in a messy, fluid, dynamic, and functional way, and classroom interaction should reflect the same, for authenticity, academic flexibility, and cultural and linguistic sensitivity to be realised. The need for the normalisation and legitimation of learners' linguistic resources and repertoires in the curriculum and the recognition of learners as emergent plurilinguals who need to employ the resources for thinking and communication cannot be overemphasised. In a witty play with words, Robertson and Graven (2020) designate English (in the South African context) as the 'language of power' whereas the learners' home languages represent 'the power of language'. It is language that mediates epistemological access, and excluding the home languages in classroom interactions is taking that power of language out of the classroom.

García et al. (2011, p. 386) posit the need to "…invert schooling structures and subvert traditional language education so as to pay attention to the singularities of students within multilingual classrooms". A de-learning and re-learning is needed to best exploit linguistic complexities in the multilingual STEM classroom. García et al. (2011, p. 384) advocate learners' active language use in dynamic relationships being the locus of control in the classroom. They also see plurilingualism being applicable where "…students' languaging is recognized and the pedagogy is dynamically centered on the singularity of the individual experiences that make up a plural". Learners utilise all their linguistic resources in task execution; informed by content, their linguistic proficiencies and preferences, and the language possibilities. Creative and spontaneous ways of validating and incorporating learners' linguistic repertoires beyond just the oral dimension and beyond code-mixing and code-switching should be celebrated if they open up epistemic access to STEM content. Knowledge production is a social process and language is a social tool so linguistic resources facilitative of social interaction are better than restrictive linguistic repertoires imposed on learners that curtail social interaction.

7.4   The Different Theoretical Tools and Approaches

The conversational and academic language distinction impels the teacher to elevate learners' languages from everyday social communicative functions to specialised academic STEM and school learning discourses. Just as the Deweyan process of reconstruction moves the learner from their present everyday experience to an organised body of knowledge, so should the linguistic dimension transit from everyday conversational language to academic specialised discourse.

The question of what is prerequisite and what is subservient to the other between language and content instruction is reflected in approaches such as content-based language teaching and language-based content instruction where one serves or drives the other. In either approach, there is a separation of the two when the ideal is their seamless blending to ensure both content and language mastery. Integrated well through meaningful communicative activities, both learners and teachers can develop bi- or multilingual competence simultaneously with content learning.

Cummins (2008) posits five cross-lingual transfer types, namely; of conceptual elements (in this case, STEM content), of metacognitive and metalinguistic strategies (like interpretation of graphic organisers), of pragmatic elements which aid meaning expression (like extra- and paralinguistic features like gestures), of specific linguistic elements (like word etymology), and of phonological awareness (sound system of language). The multidirectionality of language transfer is feasible to the extent that a sound sociolinguistic and educational environment has been created that allows for multiple languages and content to interact in complex ways.

Assessment is a thorny issue in multilingual STEM classrooms. While disentanglement of the linguistic from the socio-economic factors accounting for depressed learner achievement is onerous, Prinsloo et al. (2018) and others, largely attribute the manifest learner poor achievement (well documented in benchmark assessments) to the incongruity between learners' home language and the language of teaching and assessment. Language then becomes a key leverage point the school has control over (unlike socio-economic factors). While assessing learners in all official languages (11 for South Africa) has serious financial and logistical implications, learners should be accorded the privilege to seek clarification to assessment tasks demands in languages they are comfortable with. Such linguistic accommodations, easing language not content, would ensure that STEM disciplines test STEM content and not language; rendering them valid.

Because Africa lacks deliberate planning on bilingual education supported by theory and well-defined procedures (Clegg & Afitska, 2011), teachers should experiment with, and craft their own bilingual practices. Language is too pivotal in mediating learning to be left to chance or even one's whims and caprices. Where particular languages lend themselves to particular aspects and ways of learning, King (2018) proposes the development of a novel model of 'education for plurilingualism' where different languages are imported and utilised in education even if they may meet diverse goals depending on the levels at which they are mastered and supported. Robust research is needed to locate and uncover some intuitive or even unconscious language alternation practices that are working in multilingual contexts and theorise, describe, and popularise them. A prescriptive one-size-fits-all language treatment for the classroom linguistic diversity is not desirable as it, apart from lacking pedagogical justification, is inconsistent with language use in real life.

In terms of language alternation, code-switching, as noted earlier, accords a brief detour from the LoLT to the home language before going back to the ideal (the LoLT). Otheguy et al.'s (2015, p. 281) definition of translanguaging as "the deployment of a speaker's full linguistic repertoire without regard for watchful adherence to the socially and politically defined boundaries of named (and usually national and state) languages" is consistent with the proposal this chapter makes. The restrictive imposition of standard and acceptable linguistic resources in the classroom, which the same authors call "selective legitimation that license only linguistic features associated with powerful speakers and states" (p. 301) is counterproductive. Language in the classroom is a means and not an end in itself and whatever linguistic repertoires and resources best serve the end; epistemological access, should be deployed unreservedly. This is particularly so in the STEM classrooms where the uniqueness of individual disciplines is ideally lost as the disciplines serve as a unified knowledge body. The language hierarchies need to be lost sight of as learners deploy a cocktail of linguistic resources that best serves their understanding of content.

Translanguaging engenders multiple and fluid identities. In translanguaging, multilingual individuals systematically traverse between the languages they have proficiency in, as they engage in complex discursive practices in an integrated way. They do it seamlessly to the extent that their linguistic repertoire "…is understood as one system, rather than as a collection of discrete languages" (Childs, 2016, p. 24).

Instructional practices, even under bilingual models, have all been about learners using specific languages rather than creating plurality from individual learners' "singularities" (García, et al., 2011). Learner autonomy in language use enhances linguistic fluidity in the classroom, allowing for production of oral and literate texts in preferred languages and translated to other languages where necessary. This engenders dynamic and recursive bilingualism which recognises plurilinguals' practices as complex and interrelated. García et al. (2011, p. 384) recommend "… heteroglossic bilingual conceptualizations … in which the complex discursive practices of multilingual students, their translanguagings, are used in sense-making and in tending to the singularities in the pluralities that make up multilingual classrooms today". With translanguaging, focus is not on merely synthesising or hybridising diverse language practices (as languaging transcends a system of rules or structures), but crafting novel language practices that complexify linguistic discourses among interlocutors.

8 Conclusion

The linguistic complexity of the multilingual STEM classroom is multi-layered and nuanced than is normally appreciated. The complexity, far from being a constraining problem, is an opportunity for novel research, sober rethinking of policy, and creative practice that acknowledges the indispensability of learners' manifold linguistic resources in their learning.

Policy and practice seem to largely cohere in terms of proscription of HL to LoLT status in the Foundation Phase but inconsistent with research that recommends a longer shelf life for HLs as LoLTs beyond Foundation Phase. While policy recommends multilingualism, practice suggests monolingual practice; the latter occasioned by the shortage of African language teachers compromising teacher proficiency to navigate the linguistic diversity in the multilingual classroom. The HLs' capacitation to meaningfully contribute to classroom discourse has been questioned on account of lack of intellectualisation, lack of digital and knowledge economy presence, as well as learners' limited proficiency in the languages beyond the conversational level occasioned by the learners' premature exit from using the languages as LoLT. There is merit in having all teachers, particularly STEM and non-language teachers, educated in the art of navigating diverse languaging practices in the classroom, as classroom language use is every teacher's business.

References

Bergsten, C., & Frejd, P. (2019). Preparing pre-service mathematics teachers for STEM education: An analysis of lesson proposals. *ZDM Mathematics Education, 51,* 941–953.

Childs, M. (2016). Reflecting on translanguaging in multilingual classrooms: Harnessing the power of poetry and photography. *Educational Research for Social Change, 5,* 22–40.

Clarkson, P., & Carter, L. (2017). *Multilingual contexts: A new positioning for STEM teaching/learning.* 3R. Millar. Cognitive and Affective Aspects in Science Educatio Research: Selected Papers from the ESERA 2015 Conference (pp. 233–242). Switzerland: Springer.

Clegg, J., & Afitska, O. (2011). Teaching and learning in two languages in African classrooms. *Comparative Education, 47*(1), 61–77.

Constitution of the Republic of South Africa [South Africa]. (1996, December 10). Available at: https://www.refworld.org/docid/3ae6b5de4.html [accessed 30 July 2020].

Cummins, J. (2008). Teaching for transfer: Challenging the two solitudes assumption in Bilingual education. In J. Cummins & N. H. Hornberger (Eds.), *Encyclopedia of language and education* (pp. 65–75). Berlin: Springer.

de Oliveira, L. C. (2016). A language-based approach to content instruction (LACI) for English language learners: Examples from two elementary teachers. *International Multilingual Research Journal, 10*(3), 217–231.

Department of Education. (1996). *South African Schools Act of 1996.*

Department of Education. (1997). *Language in education policy.* https://www.education.gov.zaD ocuments/policies/LanguageEducationPolicy1997 (accessed 15 June 2020).

Fernald, A., & Weisleder, A. (2011). Early language experience is vital to developing fluency in understanding. In S. B. Neuman & D. K. Dickinson (Eds.), *Handbook of early literacy research* (Vol. 3, pp. 3–19). New York, NY: Guilford Press.

Freeman, D., & Freeman, Y. (2007). *English language learners: The essential guide.* New York, NY: Scholastic.

García, O., & Otheguy, R. (2020). Plurilingualism and translanguaging: Commonalities and divergences. *International Journal of Bilingual Education and Bilingualism, 23,* 17–35.

García, O., Sylvan, C., & Witt, D. (2011). Pedagogies and practices in multilingual classrooms: Singularities in pluralities. *The Modern Language Journal, 95*(3), 385–400.

Gunnink, H. (2014). The grammatical structure of Sowetan tsotsitaal. *Southern African Linguistics and Applied Language Studies, 32*(2), 161–171.

Hurt, H., & Betancourt, L. M. (2016). Effect of socioeconomic status disparity on child language and neural outcome: How early is early? *Pediatric Research, 79*(1–2), 148–158.

King, L. (2018). *The impact of multilingualism on global education and language learning.* Cambridge: UCLES. Lindholm-Leary.

Lefever-Davis, S., & Pearman, C. (2015). Reading, writing, and relevancy: Integrating the 3Rs into STEM. *the Open Communication Journal, 9*(1), 45–52.

Lopez, A. A., Guzman-Orth, D., & Turkan, S. (2014). *A study on the use of translanguaging to assess the content knowledge of emergent bilingual students.* Paper presented at the annual meeting of the AAAL Annual Conference, Portland.

Makoe, P., & McKinney, C. (2014). Linguistic ideologies in multilingual South African suburban schools. *Journal of Multilingual and Multicultural Development, 35*(7), 658–673.

Otheguy, R., García, O., & Reid, W. (2015). Clarifying translanguaging and deconstructing named languages: A perspective from linguistics. *Applied Linguistics Review, 6*(3), 281–307.

Prinsloo, C. H, Rogers, S. C., & Harvey, J. C. (2018). The impact of language factors on learner achievement in Science. *South African Journal of Education, 38*(1), 1–14.

Probyn, M. (2017). Languages and learning in South African classrooms: Finding common ground with north/south concerns for linguistic access, equity, and social justice in education. In P. Trifonas, T. Aravossitas (Eds.), *Handbook of research and practice in heritage language education*. Cham: Springer International Handbooks of Education.

Probyn, M. J. (2015). Pedagogical translanguaging: Bridging discourses in South African science classrooms. *Language and Education, 29*(3), 218–234.

Robertson, S. A., & Graven, M. (2020). A Mathematics teacher's response to a dilemma: 'I'm supposed to teach them in English but they don't understand.' *South African Journal of Childhood Education, 10*(1), 1–10.

Schwarzer, D., & Acosta, C. (2014). Two activities for multilingual students: Learning in monolingual classrooms. *Journal of Multilingual Education Research, 5,* 92–110.

Sibanda, J., & Baxen, J. (2016). Determining ESL learners' vocabulary needs from textbook corpus: Challenges and prospects. *Southern African Linguistics and Applied Language Studies, 34*(1), 57–70.

Sierens, S., & Avermaet, P. (2014). Language diversity in education: Evolving from multilingual education to functional multilingual learning. In D. Little, C. Leung, & P. Van Avermaet (Eds.), *Managing diversity in education: Languages, policies, pedagogies* (pp. 204–222). Bristol: Multilingual Matters.

Stoop, C. (2017). Children's rights to mother-tongue education in a multilingual world: A comparative analysis between South Africa and Germany. *PER/PELJ, 20,* 1–35.

Wedin, A., & Wessman, A. (2017). Multilingualism as policy and practices in elementary school: Powerful tools for inclusion of newly arrived pupils. *International Electronic Journal of Elementary Education, 9*(4), 873–890.

Wet, F., Louw, P. H., & Niesler, T. (2007). Human and automatic accent identification of Nguni and Sotho Black South African English. *South African Journal of Science, 103*(3–4), 159–164.

Jabulani Sibanda is a senior lecturer in Language Education, whose interests lie in second language teaching and learning broadly, and vocabulary development specifically, especially as it relates to important transitional points of the education system. His PhD in Language Education, his vast experience as a language teacher educator, and publications in the same field, equip him to write from an informed position in relation to language research policy and practice. His critical stance has seen him challenge extant research, policies and practices like the definition of word for vocabulary studies, the determination of key vocabulary for explicit instruction, and the disjuncture in research, policy and practice in language use in second language contexts. Sibanda is critical of unquestioned wholesale importation of research-based best practices in Home Language contexts to Second language contexts, and the manifest assumptions of linguistic homogeneity in either monolingual or multilingual contexts.

Approaches That Leverage Home Language in Multilingual Classrooms

Manono Poo and Hamsa Venkat

Abstract Internationally, concerns remain that approaches that leverage home languages teaching in ways that provide epistemic access remain elusive. In South Africa, the research base has been critiqued for tending to focus on either language issues or epistemic access rather than bringing these two foci together in practical and pragmatic ways for teachers. Our focus in this chapter is on exemplifying differences between two approaches to working in early grades' mathematics classes in multilingual contexts. The first approach, using translation has a long-associated literature base in substitution/code-switching. The second, and more recent approach, uses what has been described as 'translanguaging' and involves a wide range of multiple discursive practices in spatial, visual and spoken modes. In the chapter, we present and discuss the two approaches, and use grounded analysis of empirical excerpts to point to key differences between translation and translanguaging, and subtle differences within the categories dependent on how mathematical representations are traversed. This analysis leads us to suggest that the latter is the better approach compared to substitution of words and or phrases in multilingual mathematical contexts where the various registers associated with mathematical representation can also be considered as important parts of moving between languages. The analysis provides insights into how translanguaging, as an approach, allows languages to be treated as resources for learning in the early primary grades. Recommendations are made for policy makers and teacher education practitioners to ensure that mathematics teachers in the early grades have resources and skills needed to effectively teach mathematics through the medium of African languages.

Keywords Translation · Translanguaging · Conversion · Multilingualism · Mathematics teaching · Early grades

M. Poo (✉) · H. Venkat
Wits School of Education, University of the Witwatersrand,
Johannesburg, South Africa
e-mail: manono.poo@wits.ac.za

H. Venkat
e-mail: hamsa.venkatakrishnan@wits.ac.za

1   Introduction

The international literature base strongly promotes the notion of languages as resources for learning (Cummins, 2000; Wigglesworth et al., 2011). However, in postcolonial contexts like South Africa, tensions remain between home languages as languages that can support meaning-making and English as the language associated with economic and political capital (Setati, 2008). In such contexts, ways to leverage home languages teaching in ways that provide epistemic access remain elusive, with the research base critiqued for tending to focus on either language issues or epistemic access rather than bringing these two foci together in practical and pragmatic ways for teachers (Hoadley, 2012). Schleppegrell (2007) and Moschkovich (2015) agree that for learners to read, listen and speak in a mathematics class, they need to develop an academic language related to mathematics as a subject. These authors suggest that language contributes to constructing knowledge in mathematics teaching and learning. The relationship between language and mathematics is important and this importance is not made a focus in the early grades. Essien (2018) investigated the extent to which the role of language in early grade mathematics is researched in Kenya, Malawi and South Africa. He concluded that mathematics is strongly linked to language because mathematics learning involves reading, writing, listening and discussing which are all language-related activities. This author argues that while there is such a strong relationship between mathematics learning and language, there is limited research on the impact of language on the teaching and learning of mathematics in the early grades.

Compounding questions of how language can best be used to support mathematical access, there is also evidence in South Africa of gaps in primary teachers' mathematical content knowledge and pedagogic content knowledge with disruptions to coherence in early grades' instruction being relatively common (Askew et al., 2019). Access to manipulative resources also remains limited in many South African classrooms, a point that we return to later in this paper when we consider the implications of these limitations for working between languages and between mathematical registers.

In this chapter, our focus is on comparing and contrasting two approaches that have been advocated in the literature for supporting mathematics learning through attention to language in pedagogy and classrooms: translation, involving 'code-switching' between languages to support meaning-making; and translanguaging, involving flexible and extended moves between both languages and representational registers, based on the subject matter being dealt with and the sense-making needs of learners. We draw on excerpts of classroom interaction in South African Sepedi-medium mathematics classes to illustrate the overlaps and contrasts between these two approaches. We use discussion of the contrasts to build an argument in support of translanguaging as an approach that can usefully feed into policy and practice in South Africa, and elsewhere, in ways that support teachers to provide better epistemic access through bringing language together with other mathematical representations into multimodal packages that prioritize sense-making.

We share this story through the following structure. A brief outline of the South African Language in Education policy is provided to set the scene of the current landscape in early grades' mathematics. We draw on evidence on primary literacy and early mathematics performance studies to note that broad concerns remain about the efficacy of the policy. These concerns lead us into an introduction to the two approaches in the literature that we seek to focus on and contrast in this paper: translation and translanguaging. In the body of the chapter, we introduce, share and discuss empirical episodes that illustrate some of the ways in which translation and translanguaging differ, and also how the latter builds in more opportunities for incorporating language within a multisemiotic bundle of representational registers that can be marshalled together to support meaning-making. We conclude the chapter with a discussion of what these findings suggest for strengthening language policy and teacher education in relation to early grades' mathematics teaching.

2 The South African Language in Education Policy

The introduction in 1997 of the Language in Education Policy (LiEP) allowed the use of all eleven South African languages as languages of Teaching and Learning in the foundation phase (Department of Basic Education, 2010). Currently, the majority of primary schools in South Africa teach through the medium of African languages in the foundation phase and switch to English at Grade 4. These schools take in approximately three-quarters of all foundation phase learners (Spaull, 2016). The remaining minority of schools, mostly located in the historically white suburbs of urban centres and thus continuing to represent the more privileged end of the system, largely teach through the medium of English from the start (Spaull, 2013). School-based policies for how the language transition in Grade 4 should be supported vary, and occur in the context of widespread low attainment by the end of Grade 3—outlined in the next section.

3 Primary School Attainment and Mathematics Learning and Teaching Evidence

South African evidence points to poor learner performance in early literacy and mathematics. Findings of the Progress in International Reading Literacy Study (PIRLS) of (2011) showed that Grade 4 learners for whom English is an additional language performed well below expected levels even though they were given an easier assessment than their international counterparts. Specifically, for the Xitsonga, Tshivenda and Sepedi medium of instruction language groups, this report showed that 50% of learners could not read by the end of Grade 4 (Howie et al., 2012; Spaull, 2015). Looking at the pre-PIRLS nationally representative datasets of reading achievement in South Africa, Spaull (2016) noted that 58% of learners did not learn to read for meaning by the end of Grade 3.

On the early mathematics side, a range of studies have pointed to the prevalence of counting-based strategies in learner working, with limited evidence of moves beyond what van den Heuvel-Panhuizen (2001) describes as calculating-by-counting into calculation-by-structuring (more efficient strategies that make use of mathematical relationships and properties) and formal calculating (using combinations of known facts and efficient algorithms).

Spaull (2016) emphasizes the importance of language in the early years of teaching and learning but cautions researchers to show some understanding of the distinction between the language of instruction and the quality of instruction. His study showed that the literacy and numeracy performance of South African learners at Grade 3 level was lower than the grade-related expected levels of performance even before they switched to English as the language of instruction. This finding suggests that language is not the only factor that impacts learner performance in literacy and numeracy. Instead, it would appear useful to consider instruction with language as one of the factors that need to be investigated. Researchers locally and internationally generally agree that instruction in mother tongue is useful for various reasons such as improvement in learners' literacy skills which are essential for acquisition of knowledge and concepts central to reading and understanding other languages (Parry, 2000; UNESCO, 2007; Gacheche, 2010). While there is some agreement locally and internationally about the importance of mother-tongue instruction, teachers who teach through the medium of mother tongue also need the necessary support with mathematical knowledge, skills and resources to make teaching and learning meaningful (Murray, 2007).

The literature base in the South African context also points to problems relating to teaching and in particular gaps in primary teachers' mathematical content and pedagogic content knowledge (Carnoy et al., 2008; Venkat & Spaull, 2015). In their re-analysis of the teacher mathematics content knowledge aspect of the 2007 Southern African Consortium for Monitoring Education Quality (SACMEQ) dataset, Venkat and Spaull (2015) found that 79% of a nationally representative sample of Grade 6 teachers showed levels of content knowledge that was below the Grade 6/7 levels. They also found that mathematics content knowledge gaps were more marked in Quintiles 1–4 (socio-economically poorer) schools than in Quintile 5 (socio-economically wealthier) schools. Ensor et al.'s (2009) empirical analyses pointed to limited awareness of mathematical progression towards more efficient methods among early grades' teachers. Once again, much of this research base suggests that locating attention to language within a mathematical focus is likely to be necessary.

4 Early Grades' Mathematics Learning: Moving Between Representational Registers

Successful early mathematical learning is often described as involving confidence and competence with representational and linguistic repertoires. The importance of moving between mathematical representations is widely acknowledged as a central pillar of mathematical working (Doerr & Lesh, 2011; Lesh & Lehrer, 2003). In considering the representational repertoires that are critical within early mathematical learning, Haylock and Manning (2014) have pointed to helping children to make connections between language (both written and spoken), concrete experiences, pictures and symbols as critical. Duval (2006) has emphasized that construction of connections between representational registers has to be made explicit for learners to develop rich understandings of mathematical ideas.

While moving between mathematical representations is seen as central to mathematical working, it is also acknowledged to be difficult. Several researchers have commented that moving between representations involves recognizing the essential structural relations in one register and then translating this structure into another register, with awareness of what has to be retained and what can change in this move (Noble et al., 2001; Ainsworth, 1999). For example, in the move from iconic images of apples packed into bags of four into a ratio table for working out the number of bags needed to pack 48 apples, children learn that all the intermediate numbers of bags and apples may not be necessary to get to the result, and that a similar table can be used to represent the number of tables with 48 legs. Duval's (2006) empirical studies indicate that students generally find it easier to make moves within a particular representational register (which he describes as 'treatment' moves) than to move between registers (described as 'conversion' moves).

Several writers in early years' mathematics describe the key registers that children encounter in terms of: oral language, concrete working with manipulative objects, iconic or indexical diagrams and symbolic forms (e.g. Haylock & Manning, 2014). In Duval's (2006) terms, moves between any registers count as conversion moves, but in the first author's doctoral study, we found it useful to introduce a caveat to this interpretation. Specifically, while some moves between registers involve awareness or reconstruction of what English (1993, p. 8) describes as 'structural similarities', other moves between registers do not require this. For example, if a child is able to represent an additive word problem in a part-part-whole diagram, we might say that the child has internalized the relations between quantities in the problem and can present this relation in another register. However, in episodes involving an oral 'reading' of a symbolic representation, or vice versa, the symbolic writing down of a verbal representation, the connecting of symbolic and linguistic representations here does not require any understanding of the mathematical structure of one representation that needs to be converted into the register of the other representation. For example, with $7 + 2$ written on the board, the teacher says: 'I have written seven plus two'. Oral and symbolic registers are linked here but no awareness of structure is required—there is simply a restatement. This is even more marked when what is being read or written relates to previously introduced, rather than new, ideas. This is not to say that learning to 'read' symbolic representations or write down verbal representations in symbolic registers is unimportant. On the contrary, it is critical to learning the 'language' of mathematics. But at the same time, this learning does not require any awareness of the structure of a representation that needs to be carried across a conversion move. Essentially, it is a different kind of conversion move, and one that is more akin to translating between languages than to converting structurally between representations.

A further aspect of using multiple representations that has been highlighted in the South African literature (Ensor et al., 2009) is that moves into more formal, or abstract, mathematical representations provides a way of 'specializing', or moving into the specialized discourse of mathematics as a discipline, with firm anchors in meanings based in everyday language and concrete actions.

Taken together, the literature base on multiple representations points to instruction needing to focus on two aspects: firstly, supporting children to make connections between representations by juxtaposing them and explaining and elaborating the linkages; and secondly, connecting situations described in everyday language and involving concrete actions with increasingly formal mathematical representations.

5 Early Grades' Mathematics Learning: Moving Between Languages

While language features as one vertex in Haylock and Manning's (2014) identification of representational registers, much of the work on multiple representations has been developed in monolingual or dominant single language settings. The emphasis of the work, therefore, has been on ways to support children to move between mathematical representations without the need to consider a terrain where language development is simultaneously fragile and in the throes of multilingual transition.

In bringing multilingual moves within the scope of interest in moves between representations, we found it useful to add moves between languages as one aspect of moves between representations. This was particularly useful given the language policy context that involves a move to English (in most schools in South Africa) as the medium for mathematical instruction in Grade 4, with Grade 3 frequently described by teachers as the school year in which children need to be supported to 'transition' from home languages into English, if they are to successfully access mathematics in the later grades. Moving between languages has been widely described as an important facilitator of sense-making for learners in the urban multilingual settings of our work. The literature on language in mathematics education offers, as noted already, translation and translanguaging moves between languages as routes that can support sense-making.

Translation/substitution is a practice used by multilingual individuals to move between two languages. This practice entails the substitution of one word or phrase in one language with a phrase/word in another language (Baker, 2011; Childs, 2016). Adler (1999) defines this move between languages as code-switching. She defines code-switching as an alternation in use of more than one language in a single speech act. It is often responsive and unplanned. In contrast, translanguaging is often described as centrally involving the purposeful alternation of languages in spoken and written forms (García & Wei, 2014). Childs (2016) notes that while translation, code-switching and translanguaging are all multilingual practices observable in multilingual classrooms, code-switching and translation are 'responsive' practices, used to respond quickly and constructively in the moment to learners' responses. In contrast, translanguaging is viewed as a planned teaching strategy. We use the term 'planned' in this chapter not in the sense of activities for the lesson decided in advance, but rather, in the sense that there is intentional attention to broad working with multiple representations across language and mathematics with the mathematical topic in focus at the heart of these decisions.

Beyond the 'in the moment' versus 'planned' distinction between translation and translanguaging, definitions of translanguaging point to a remit that goes beyond a primary, or even sole, focus on oral language. Instead, translanguaging is described as including spoken and written registers and a range of other cognitive and semiotic resources to make meaning and sense (Baker, 2011). Translanguaging therefore takes on the importance of moves between mathematical representations as well as moves between languages. In the context of mathematics, translanguaging relates to multimodal working that emphasizes the use of language and multiple modes to make meaning (Joutsenlahti & Kulju, 2017). Translanguaging therefore involves a systematic use of language and registers that go beyond simple substitution of one representation with another. Instead, we refer to a systematic awareness of a variety of ways in which teachers use a language and other mathematical representations to communicate meanings about mathematical concepts. Moschkovich's (1999) exemplification of 'discourse' focused approaches provide useful instances of this kind of broadened repertoire of interconnections between moves between languages and other registers. Our context though, with multiple languages of instruction in the national terrain, differs in some ways from Moschkovich's context, with a less immediate need to learn English. This leads to broader attention to how moves between languages and between representations interplay in our context in ways that may help us to understand differences in mathematical attainment. Our approach also contrasts with the relatively small literature base on translanguaging in mathematics education, where more attention has been paid to multirepresentational moves in largely monolingual settings (e.g. Joutsenlahti & Kulju, 2017). In this literature base, attention turns, understandably, to exploring links between natural and mathematical language.

The literature base provides examples that helped us to think about the differences between translation/translanguaging connected with treatment/conversion moves. For example, the excerpts of instruction that we use to illustrate the differences between translation and translanguaging in this chapter are drawn from Sepedi-medium classrooms, Sepedi being one of the eleven official South African languages. An important point to note about Sepedi (which also holds for several of the indigenous South African languages) is that it has a very logical and transparent system of number naming. While this transparency is not the focus in this chapter, Mdluli (2017) has analysed in earlier writing, the lack of explicit reference in instruction, to the logic of the naming structure in relation to the quantities being represented. In this chapter, Sepedi extracts are presented verbatim with our translations, offered in bracketed italics, setting out the literal English translation and the common English naming of numerical quantities.

Our aim in this chapter is to illustrate key aspects of the differences between translation and translanguaging. In discussing these excerpts, we note the ways in which translanguaging moves between the Sepedi and English languages frequently involve explicit links between spoken *and* written forms in Sepedi and/or in English. Additionally, in line with the more inclusive attention in translanguaging to a range of semiotic modes, the translanguaging excerpts we focus on frequently include reference to a range of mathematical representational registers. Our analysis of the excerpts allows for a building of an argument for why we believe translanguaging, with its considered incorporation of moves between representations, provides better access to meaning-making in early grade mathematics than translation.

The literature on bi- and multilingualism in education does not include a hierarchy of levels when referring to translation and translanguaging. In this chapter, we conclude with a clear distinction between translation and translanguaging moves that place these pedagogic moves in a hierarchical relation to each other, in ways that link with Duval's hierarchy between treatment and conversion moves. The translation/translanguaging hierarchy is justifiable theoretically, on the basis that translanguaging moves offer more explicit pointers across languages and representational registers to alternative ways of expressing mathematical ideas. In this chapter, we offer empirical support to this claim through the analysis of classroom extracts showing teacher moves between mathematical and multilingual representations.

6 Data Sources and Analytical Methodology

The instructional excerpts that we draw on in this chapter come from videotaped lesson observations of lessons taught by two Grade 3 teachers, Nkele and Mirriam (pseudonyms) in two township primary schools in the East of Johannesburg in South Africa. These two teachers featured in the first author's doctoral study looking at teachers' work with language in different language settings (Poo, 2020). Both schools are large and serve disadvantaged learner populations, with different classes in each of the foundation phase grades constituted on the basis of pupils' home languages. These two teachers were chosen on the basis that they taught mathematics in Sepedi-medium classrooms using combinations of English and Sepedi in their instruction. Sepedi was Nkele and Mirriam's first language, but both of them could also speak English and at least one of the other South African languages.

In the broader study (Poo, 2020), verbatim records of all instructional talk, pupil questions and responses and how representational modes featured in each lesson were produced. Gestures and shifts in tone/inflection were included in the lesson transcripts. Our selection of excerpts for this chapter from the broader dataset was guided by the literature-based descriptions of translation and translanguaging. Specifically, we aimed to study differences in the ways translation and translanguaging played out, and to consider the potential implications of these different enactments for sense-making and learning. As noted already, in our presentation of illustrative excerpts in this chapter, we present talk in verbatim format, with literal and/or everyday translations into English in italicised square brackets for the Sepedi tracts. Curly brackets are used for paraphrased descriptions of sections of the enactment. We considered translation, coded as ML1, and translanguaging, coded as ML2, as instances where the teacher moved between English and Sepedi. Direct substitution, of words or phrases, was considered as a translation move. Where moves between languages involved elaborations based in either language or in other multimodal resources, this was interpreted as translanguaging. These moves between languages were juxtaposed in our analysis with the ways in which teachers' moves between mathematical registers, with treatment moves coded as MM1 and conversion moves, in Duval's terms, coded as MM2.

7 Illustrative Excerpts: Commentaries and Analysis

We begin this section with two illustrative excerpts of instruction (see Fig. 1). These excerpts overlap in their incorporation of translation moves between languages, but contrast in the ways in which moves between mathematical representations are worked with, and helped us to think about the distinction pointed out earlier about less and more mathematically oriented conversion moves.

In the first excerpt, the teacher acknowledged the answer offered by learners, and repeated it in English and then in Sepedi. The teacher then moved the orally offered number into symbolic form by writing in '12' into the number sentence on the board. Translation moves between English and Sepedi and the restatement type of conversion moves between symbolic and oral language number forms we described earlier therefore occurred in the first excerpt. Excerpt 2, drawn from Mirriam's class, is similar in many ways to Excerpt 1: a translation move is evident in the teacher first stating the number name orally in Sepedi and then in English. There is also a pointing to the symbolic numeral representation of 376 following its oral presentation, but here, there is a small, but important distinction in that 376 is featured within a 100-number-chart representation. In South Africa, Ensor et al. (2009) have noted the widespread prevalence of unit counted versions of number, and the limited inclusion of symbolic number system-based representations of number. While the base ten structure is not referred to in the instructional talk, the inclusion of this artifact expands the working here a little further into the conversion move terrain than Excerpt 1. Excerpt 2, therefore, offers a marginally widened repertoire of work with conversion moves in comparison to Nkele's teaching in the first excerpt.

Excerpt 1: Nkele ML1 & MM2 (restatement move)	Excerpt 2: Mirriam ML1 and MM2 {Learners have the 301-400 number charts on their desks.}

Within a section of whole class teaching of a sequence of examples of missing addend tasks with total 20, the teacher writes: 8 + ☐ =20 on the board. The following interaction then ensues:
 	
   T: {Pointing to 8 on the board:} Re ka hlakantšha nomoro e le eng go bona masomepedi? *[What can we add to this number to make twenty?].* We need to add something to it so that we make masome pedi *[tens two/twenty].* {Calls out a learner's name for a response.}
 	
  L1:   Lesome pedi *[ten two/twelve].* {Said immediately without counting.}
 	
  T: We can add eight and twelve, lesome pedi. *[Ten two/twelve].* {T fills in 12 in number sentence}. Do you understand how you counted to get masome pedi *[tens two/twenty]*?
 	
  Ls: Yes.
	T: Put your fingers on makgolo tharo masome šupa tshela *[hundreds three, tens seven, six/three hundred and seventy-six/* three hundred and seventy-six].
	
	{Pause.}
	
	T: Point to three hundred and seventy-six. Say the number and point.
	
	T: Yes, makgolo tharo- masome šupa tshela *[hundreds three, tens seven, six/ three hundred and seventy-six].* {Points to 376 on the large number chart on the board}

Fig. 1 Two different translation moves

The subsequent two excerpts, both drawn from Nkele's class teaching illustrate a different type of language move to the translation move observed in Excerpts 1 and 2 (see Fig. 2). As with the two excerpts above, each of these excerpts is accompanied by a different type of conversion move pointed out in the two previous excerpts.

Excerpts 3 and 4, in our analysis, involve instances of the intentional use of moves between languages for meaning-making. Excerpt 3 begins with the oral 'reading out' of the $3 + ☐ = 20$ number sentence written on the board. A basic move between symbolic and oral language representations, therefore, occurs early in this excerpt similar to that observed in Excerpt 1. In the context of the incorrect offer of 70, the teacher incorporates an intentional use of moves between languages to highlight the difference between the number names for 17 and 70 in Sepedi. The more marked difference in number names in Sepedi compared to English is used here to support awareness of the number distinctions, and a correct answer is subsequently offered. The teacher goes on to contrast the 'lesome' part ('one ten') with the 'masome' part

Excerpt 3: Nkele ML2 & MM2 (restatement move)	Excerpt 4: Nkele ML2 & MM2
Later, within the same sequence of examples in focus above, Nkele offers this instructional interaction when dealing with the task: 3 +□ =20, written on the board:	{Teacher takes out three number cards with numerals 100, 30 and 5 written on the cards from a container on her table. She calls three learners to the front and gives each learner one of these number cards: 30 to Jabulani, 5 to Relebogile and 100 to Thabo. She rearranges the learners from the left to the right as 100, 30 and 5.}
T: {Pointing to 3+□ =20}: What can we add to 3 to make twenty? What number is that? {L1 name}	
	T: Lekgolo- masometharo- hlano *[Hundred, tens three, five/one hundred and thirty-five]*. What number is this? {Moves her hand to point to across all three children}
L1: Seventy.	
T: What? {Teacher pauses, looks surprised.} Say that again.	Ls: Lekgolo masometharo- hlano. *[Hundred, tens three, five/one hundred and thirty-five.]*
L1: Seventy.	
T: Say the number in Sepedi so you can hear what you are talking about.	T: All I wanted to explain to you with this activity is the place value of numbers. Place value of numbers. {Points to 3 in the number 135 on the board}. This is not just 3. Ke masometharo, *['It is tens three/thirty']*, {pointing to 3 in the number 135 on the board}. It stands for 3 tens. {Holds Jabulani who has the number card written masome/tens}. The one stands for hundred. Ke lekgolo. {Points to 1 in 135 and moves to hold Thabo who has the hundred/ makgolo name tag}. It is not just one. It is one hundred. Five stands for units. It is five. Do we understand each other?
L1: Lesome supa. *[Ten seven/seventeen.]*	
T: Seventeen. Not seventy. If you say seventy then you are talking about masome šupa. *[Tens seven/seventy]*. There is a difference between masome šupa *[tens seven/seventy]* and lesome -šupa *[ten seven/seventeen]*. {T writes 17 and 70 on the board and points to each number as she says them}. Lesome-šupa *[ten seven]* is one ten. Masome- šupa *[tens seven/seventy]* is many tens, seven tens. Do you see it?	
Ls: Yes ma'am	
T. {Pointing to 3 in the number sentence on the board}. Now count from this number to twenty.	

Fig. 2 Two different translanguaging moves

(many tens), through her offer of translations into English of both of these parts. Subsequently, a unit counting on from 3 to 20 is enacted that also produces '17' as the answer. While concrete counting actions, as a further representational form, are incorporated here, the actions remain at the level of unit counting, with no reference to using the base ten benchmarks that form a transparent part of the ways in which numbers are named in Sepedi.

In Excerpt 4, we see fluid moves between languages with elaborations both in terms of language and moving between oral and symbolic mathematical registers: when the digit 3 in 135 is pointed to on the board, this is accompanied by an explanation that emphasizes: 'This is not just 3. Ke masometharo, ['It is tens three/thirty'], {pointing to 3 in the number 135 on the board}. It stands for 3 tens. As with Excerpt 2 above though, what is added here is a reference to how the numbers are structured within the base ten place value system.

In this explanation, the teacher does not only substitute a word or phrase with a corresponding number word or phrase in another language or moves between the basic oral and symbolic representations. She makes connections between learners representing numbers in different 'positions', the digits on number cards and the language of the number system using Sepedi and English as she explains how the numbers are structured in the decimal system. This multimodal way of moving between the two languages, involving pre-prepared resources as in Excerpt 2, suggests a planned instruction that goes beyond substitution. In the context of this chapter, we refer to such moves as translanguaging moves which, as noted earlier, are more akin to conversion moves in Duval's (2006) theory of mathematical moves between registers.

8 Discussion

Across the four excerpts contrasted in this chapter, we have pointed to quite subtle differences in the types of multilingual and mathematical moves observed in the teaching of early number in the two Grade 3 classrooms sampled above. The first two contrastive excerpts show what translation moves can look like and also show that translation moves involve moves between the symbolic and the oral modes of representation. While a switch between the two languages (translation move) observed in these first two excerpts is seen as a positive way to respond or provide feedback in the context where the teacher thinks that learners may not know or understand a particular word or phrase, this move tends not to make explicit the mathematics and mathematical structures embedded in the task. As observed in the first excerpt, this type of language move, while accompanied by a basic conversion move (involving moves between the symbolic and the oral representation), does not necessarily elaborate learners' understanding of mathematical concepts or ideas beyond their access to a word or phrase in another language as demonstrated in Excerpt 1. In the second excerpt, the same translation move is accompanied by a conversion move which is different from a move between the oral and the symbolic noted in Excerpt 1. The conversion move observed in the second excerpt involves a mathematical move from the oral to the symbolic number-based mode of representation. In the second excerpt, a translation move is accompanied by a move between a number-chart resource that offers access to base ten number structure and the oral language. This is a different type of conversion move because it involves moves between two very different mathematical registers. Duval's work (2006) suggests that conversion moves that start to make reference to underlying structural relations of the mathematical objects in focus (numbers primarily, in this chapter) are more complex than basic moves between representations. The difference here is subtle, and we acknowledge that the appropriateness of the move is always contingent on learner understandings: if a learner already has a strong conceptual sense of a mathematical object in the home language, then a translation move will suffice. However, as noted already, the South African evidence base details lags in learning and points to problems within instruction in terms of providing improved access to mathematical ideas. Given this evidence, our position is that the inclusion of alternative registers that elaborate the structure of the mathematical idea in focus is more helpful than sole reliance on translation moves.

Excerpts 3 and 4 show a different kind of language move to the translation and conversion moves observed in Excerpts 1 and 2. In these excerpts drawn from Nkele's class, we see translanguaging moves that elaborate on the mathematical concept of place value and number. The moves shown in Excerpts 3 and 4 exhibit and go beyond something the teacher does when she feels that learners lack a word, or a phrase needed to express them. In the third example, the translanguaging move extends beyond a basic move between the symbolic and the oral representation to the intentional use of Sepedi, rather than English to highlight a mathematically important distinction that is more marked in the former spoken form than in the latter. In parallel with the second excerpt, the fourth excerpt couples translanguaging moves with moves that include decomposed place value number cards, again, a resource pointing to number structure.

Across the two sets of excerpts, we infer that conversion moves that simply restate, in oral or written form, a word or phrase, offer similar openings to translation moves: another form of the same idea is linked with the original, but we do not learn more about the idea in this kind of move. In contrast, mathematical conversion moves that go beyond oral or written restatement of a word or phrase offer elaboration in the same ways that translanguaging moves can. Thus, we conclude that translanguaging moves are more useful to use in mathematics teaching because of the elaborated access to mathematical ideas that they offer than translation moves. This points us towards some important policy and practice level implications emanating from these findings. but are not an easy endeavour because they require moves between different modalities and connections between these modes to be made explicit.

9 Policy and Practice Implications

Our findings suggest that both translanguaging moves and conversion moves that go beyond simple restatement are particularly important for broadening access to mathematical ideas. However, in thinking about the implications for policy and practice of this finding, key contextual factors need to be taken into account. Firstly, we acknowledge that richer networks are possible than seen in the translanguaging excerpts presented in this chapter. Specifically, there is limited inclusion of diagrammatic or concrete resources that can help to illustrate relative magnitudes across place values in the base ten system, for example Dienes blocks or place value counters. This absence though, is typical of the limited range of resources that continues to be the norm in many South African classrooms as discussed in Spaull's (2013) work. In our work in schools, we have seen that 100-square type number charts are the most commonly available representation relating to base ten structure. Dienes blocks and place value cards are much rarer. The paucity of resources makes the move between oral and symbolic representations more common in such contexts than in more advantaged contexts where the availability of a range of resources can commonly be taken for granted.

The second point follows, in many ways, from the first. With limited access to physical resources, the burden on language to do the explanatory work, to incorporate reasoning and justification, is greater. This makes small and subtle differences in the ways in which moves between languages and between mathematical registers are used. This chapter shows that in mathematics classrooms, translanguaging and translation moves cannot be placed on the same level when debating the access to meaning-making. Translanguaging moves involve deeper understanding than translation as translanguaging moves go beyond substitution of words or phrases. Translanguaging involves more intentional and/or more elaborated processing and meaning-making as shown in the four classroom examples presented earlier in this chapter. This observation is supported by the work of García and Wei (2014) who argue that translanguaging carries greater potential for meaning-making.

We have found these insights useful to share in our work in pre- and in-service teacher education. The kinds of contrastive examples that we have presented in this paper provide useful exemplars for discussion in pre-service teacher education. We are in the process of adding in excerpts that include the richer networks of resources such as place value cards which are relatively easy and cheap to create alongside translanguaging in our pre-service teacher education programme materials. These contrastive exemplars are intended to help teachers to think about their own instruction in ways that allow a focus on language use in classrooms in instructional terms, as well as in political terms. Simultaneously, it helps us to address some of the gaps identified in primary teachers' mathematical content and pedagogic content knowledge. Our hope is that through this kind of feeding into teacher education, we can support student teachers in the teaching of mathematics in the early grades to adopt and use translanguaging moves as a resource that enables a broadened and increased access to fundamental mathematics ideas. We are feeding in our examples of the range of mathematical work that is possible in the context of translanguaging into discussions with other universities and with curriculum writers and policy makers. Through this, we hope to feed into policy development that can include attention to and enhance teachers' use of translanguaging approaches and multimodal languaging in the Language and Education policy in ways that deliver better on the promise of language as a resource in multilingual mathematics classrooms that can support more equitable attainment outcomes in the subject.

References

Adler, J. (1999). The dilemma of transparency: seeing and seeing through talk in the mathematics classroom. *Journal for Research in Mathematics Education, 30*(1), 47–64.

Ainsworth, S. (1999). The functions of multiple representations. *Computers & Education, 33*(2–3), 131–152. https://doi.org/10.1016/S0360-1315(99)00029-9.

Askew, M., Venkat, H., Abdulhamid, L., Mathews, C., Morrison, S., Ramdhany, V., & Tshesane, H. (2019). Teaching for structure and generality: Assessing changes in teachers mediating primary mathematics. In M. Graven, H. Venkat, A. Essien, & P. Vale (Eds.), *Proceedings of the 43rd Conference of the International Group for the Psychology of Mathematics Education* (Vol. 2, pp. 41–48). Pretoria: PME.

Baker, C. (2011). *Foundations of bilingual education and bilingualism* (5th ed). Clevedon: Multilingual Matters.

Carnoy, M., Chisholm, L., & Baloyi, H. (2008). Uprooting bad mathematical achievement: Pilot study into roots of problems. *HSRC Review, 6*(2), 13–14.

Childs, M. (2016). Reflecting on translanguaging in multilingual classrooms: Harnessing the power of poetry and photography. *Educational Research for Social Change, 5*(1), 22–40. https://doi.org/10.17159/2221-4070/2016/v5i1a2.

Cummins, J. (2000). *Language power and pedagogy: Bilingual children in the crossfire*. Clevedon, England: Multilingual Matters.

Department of Basic Education. (2010). *The status of the language of learning and teaching (LOLT) in South African public schools: A quantitative overview*. Pretoria, South Africa: Department of Basic Education.

Doerr, H. M., & Lesh, R. (2011). Models and modelling perspectives on teaching and learning mathematics in the twenty-first century. In G. Kaiser, W. Blum, R. Borromeo Ferri, & G. Stillman (Eds.), *Trends in teaching and learning of mathematical modelling* (ICTMA 14, pp. 247–268). Dordrecht: Springer. https://doi.org/10.1007/978-94-007-0910-2_26.

Duval, R. (1993). Registre de représentations sémiotique et fonctionnement cognitif de la pénse. Annales de Didactique et de Sciences Cognitives, 5, 37–65.

Duval, R. (2006). A cognitive analysis of problems of comprehension in a learning of mathematics educational studies in mathematics, Vol. 61, No. 1/2, Semiotic perspectives in mathematics education: A PME special issue (2006). *Educational Studies in Mathematics, 61*(1), 103–131. https://doi.org/10.1007/sl0649-006-0400-z.

Ensor, P., Hoadley, U., Heather, J., Kühne, C., Schmitt, E., Lombard, A., & van den Heuvel-Panhuizen, M. (2009). Specialising pedagogic text and time in foundation phase numeracy classrooms. *Journal of Education, 47*, 5–30.

Essien, A. (2018). The role of language in the teaching and learning of early grade mathematics: An 11-year account of research in Kenya, Malawi and South Africa. *African Journal of Research in Mathematics, Science and Technology Education, 22*(1), 48–59.

Gacheche, K. (2010). Challenges in implementing a mother tongue-based language in education policy: Policy and practice in Kenya. *POLIS Journal, 4*, 1–45.

García, O., & Wei, L. (2014). *Translanguaging: Language, bilingualism and education*. London, UK: Palgrave Macmillan Pivot.

Haylock, D., & Manning, R. (2014). *Mathematics explained for primary teachers* (5th ed.). London: Sage.

Hoadley, U. (2012). What do we know about teaching and learning in South African primary schools? *Education as Change, 16*(2), 187–202. https://doi.org/10.1080/16823206.2012.745725.

Howie, S., Van Staden, S., Tshele, M., Dowse, C., & Zimmerman, L. (2012). *PIRLS 2011: Progress in International Reading Literacy Study 2011: South African children's reading literacy achievement: Summary report*. Summary Report, Centre for Evaluation and Assessment, University of Pretoria, Pretoria.

Joutsenlahti, J., & Kulju, P. (2017). Multimodal languaging as a pedagogical model—A case study of the concept division in school mathematics. *Education Journal of Science, 7*, 9.

Lesh, R., & Lehrer, R. (2003). Models and modelling perspectives on the development of students and teachers. *Mathematical Thinking and Learning, 5*(2–3), 109–129. https://doi.org/10.1080/10986065.2003.9679996.

Mdluli, M. (2017). Language in early number learning in South Africa: Linking transparency and explicitness. In M. Graven & H. Venkat (Eds.), *Improving primary mathematics education, teaching and learning: Palgrave studies in excellence and equity in global education* (pp. 115–128). London: Macmillan.

Moschkovich, J. N. (1999). Supporting the participation of English language learners in mathematical discussions. *For the Learning of Mathematics, 19*(1), 11–19.

Moschkovich, J. N. (2015). Academic literacy in mathematics for English learners. *The Journal of Mathematical Behavior, 40*, 43–62.

Murray, C. (2007). Reflections on the question of mother tongue instruction in Namibia. *NAWA Journal of Language and Communication*, 69–77, viewed 11 April 2018. http://ir/wst.na/bitstream/hanlde/0628/120/Murray.

Noble, T., Nemirovsky, R., Wright, T., & Tierney, C. (2001). Experiencing change: The Mathematics of change in multiple environments. *Journal for Research in Mathematics Education, 32*(1), 85–108. https://doi.org/10.2307/749622.

Parry, O. (2000). *Male underachievement in high school education in Jamaica, Barbados and St Vincent and the Grenadines*. Kingston: Canoe Press.

Poo, M. (2020). *Exploring evaluative criteria and modes of representations in early number teaching across English and Sepedi medium classrooms*. Doctoral thesis, University of the Witwatersrand, Johannesburg, South Africa.

Schleppegrell, M. (2007). The linguistic challenges of mathematics teaching and learning: A review. *Reading Writing Quarterly, 23*, 139–159. https://doi.org/10.1080/10573560601158461.

Setati, M. (2008). Access to mathematics versus access to the language of power: The struggle in multilingual classrooms. *South African Journal of Education, 28*(1), 103–116.

Spaull, N. (2016). Disentangling the language effect in South African schools: Measuring the impact of language of assessment in grade 3 literacy and numeracy. *South African Journal of Childhood Education, 6*(1). a 477, 1–20.

Spaull, N. (2015). Schooling in South Africa. How low-quality education becomes a poverty trap. *South African Child Gauge, 12*, 34–41.

Spaull, N. (2013). Poverty & privilege: Primary school inequality in South Africa. *International Journal of Educational Development, 33*(5), 436–447.

UNESCO. (2007). *Mother tongue-based literacy programmes: Case studies of good practice in Asia*. Bangkok: UNESCO.

van den Heuvel-Panhuizen, M. (2001). Realistic mathematics education in the Netherlands. In J. Anghileri (Ed.), *Principles and practice in arithmetic teaching Innovative approaches for the primary classroom* (pp. 49–63). Buckingham, UK: Open University Press.

Venkat, H., & Spaull, N. (2015). What do we know about primary teachers' mathematical content knowledge in South Africa? An analysis of SACMEQ 2007. *International Journal of Educational Development, 41*, 121–130.

Wigglesworth, G., Simpson, J., & Loakes, D. (2011). NAPLAN language assessments for indigenous students in remote communities: Issues and problems. *Australian Review of Applied Linguistics, 34*(3), 320–343.

Manono Poo holds a Ph.D. from the University of the Witwatersrand. Her Ph.D. focused on Primary Mathematics Education. She trained as a Foundation Phase teacher and practiced teaching for five years prior to undertaking tertiary education in Bachelor of Arts with Education as a major. She holds a postgraduate degree in Personal and Professional Leadership and a Master's degree in educational management which she obtained from the University of Johannesburg, South Africa. She has coauthored some articles with colleagues within the Wits School of Education and has published a book chapter on her own.

Hamsa Venkat is a Professor of Mathematics Education and holds the South African Numeracy Chair at the University of the Witwatersrand in Johannesburg—now in its second 5-year phase of research and development in primary mathematics. She leads a team of academics and post-doctoral and postgraduate students, all involved in studying and improving primary mathematics teaching and learning in government primary schools serving disadvantaged students. Her work in South Africa has been in the areas of Mathematical Literacy and Primary Mathematics. Prior to this, Hamsa was based in England, working initially as a high-school mathematics teacher in London comprehensive schools, before moving into teacher education at the Institute of Education and research in mathematics education at King's College London. She has published widely, across articles and books, and her research work continues to feed into provincial and national policy initiatives.

Creating Dialogues in Whole Class Teaching in Multilingual Classrooms: Language Practices and Policy Imperatives

Audrey Msimanga

Abstract Teaching science or indeed any subject in a language that learners are not proficient in is difficult even for the best of teachers. In South Africa, the situation is compounded by various contextual issues including a long tradition of the dominance of transmission methods and teacher talk. The result is poor achievement in science as learners simply memorise and regurgitate concepts in examinations. Yet, one of the guiding principles of South Africa's National Curriculum Statement is to achieve "Active and critical learning: encouraging an active and critical approach to learning, rather than rote and uncritical learning of given truths" (Department of Education, The National Curriculum Statement (NCR): Curriculum and Assessment Policy Statement Grades 10–12 Physical Sciences, Department of Education, Pretoria, p. 4, 2012). Thus, the curriculum explicitly discourages uncritical learning. Recent research has explored small group work as a potential strategy to promote active learner engagement. However, the uptake of group work remains low. Teachers are not confident in managing group work while teaching the content-heavy curriculum to often very large classes in the challenging contexts of multilingualism. In this chapter, I draw on (Mortimer & Scott, in *Meaning making in secondary science classrooms*. McGraw-Hill Education, Berkshire, UK, 2003) framework to illustrate the potential for whole class teaching to create dialogic discourse that enables the active learner engagement anticipated in the South African curriculum. I discuss some of the tensions that such an approach raises in the current South African language policy context, in particular the implications for leveraging the linguistic resources of the classroom to optimise learner participation.

Keywords Dialogic discourse · Whole class teaching · Multilingual policy · Monolingual science classrooms

A. Msimanga (✉)
School of Education, Sol Plaatje University, Kimberley, South Africa
e-mail: audrey.msimanga@wits.ac.za

School of Education, University of the
Witwatersrand, Johannesburg, South Africa

© The Author(s), under exclusive license to Springer Nature Switzerland AG 2021
A. A. Essien and A. Msimanga (eds.), *Multilingual Education Yearbook 2021*, Multilingual Education Yearbook,
https://doi.org/10.1007/978-3-030-72009-4_4

1 Introduction

One of the guiding principles of South Africa's National Curriculum Statement is to achieve learner active involvement in their own learning. This desire has been articulated in different ways both in the curriculum documents and the various forums for its implementation including teacher education, professional development interventions as well as research communities. In fact, the National Curriculum Statement states "Active and critical learning: encouraging an active and critical approach to learning, rather than rote and uncritical learning of given truths" (Department of Education, 2012, p. 4). Thus, the curriculum explicitly discourages rote learning and transmission methods (uncritical learning). Education research in South Africa has addressed this principle through the adoption of learner-centred teaching approaches. In Mathematics and Science education, research has focused on small group work as a preferred method towards learner-centredness. Group work is deemed suitable to address curriculum goals by providing "support for the construction of ... meaning ..., since it allows more time and space for ... talk and activity" (Brodie, 2000, p. 9). Group work became the focus of much research that aims to address the long history of traditional teacher-centred methods and teaching to the test. Yet, literature on classroom-based research in South Africa like elsewhere in the world reports the persistence of traditional transmission methods and prevalence of teacher talk. Small group work remains a challenge for South African teachers not only because it requires specific skills to plan the tasks and manage the group work, but also because they find it difficult to sequence and time the lesson progression to be able to cover the rather content-heavy syllabus adequately. Also, learners have been observed to shift to using their home languages when they are placed in small groups. Teachers worry that learners may not stay on task if allowed to work on their own in small groups, especially when they engage in their home languages. The large class sizes and overcrowded classrooms only exacerbate the problem. Generally, there are many genuine reasons why teachers find it difficult to use small group discussions in science teaching. Hence the persistence of whole class teaching in many South African science classrooms. Thus, in spite of the many interventions in the past two decades to change pedagogical practices to learner-centeredness, there is still very little learner talk and activity in South African classrooms and science classrooms are not different. Classroom interaction continues to be largely through whole class mode characterised by recitation and memorisation.

While small group work is espoused as the best approach to get learners talking and transform classrooms to learner-centred, the classroom context in South Africa does not seem to be conducive to small group activities. Meanwhile, research elsewhere shows that whole class teaching has the potential to create the kind of learner interaction anticipated in the South African curriculum. According to Lyle (2008), whole class discussions can develop into collaborative dialogic talk creating dialogic rather than the current prevalent monologic discourses. Dialogic talk according to Lyle creates spaces for learner voice, allowing learners to ask questions, explore each other's ideas and change their minds. In many ways, this is what is anticipated in the South African curriculum and small group interactions. However, in multilingual classrooms, the challenge with achieving such dialogic classroom interaction is not just the infrastructure or teacher preparation. The challenge is also language. English is the preferred LOLT for many in South Africa and yet it is a second, third or even fourth language for the majority. Thus, many learn science in a language that they are not sufficiently proficient in. It has been established that English second language learners (ESLs) struggle to build registers for the language of instruction (Lyle, 2008; Milligan & Tikly, 2016). Their first hurdle is just to be able to talk in English. And only then can they make sense of the content. For science teachers of ESLs then the task is both to enable talk and then to mediate the talk for meaningful science learning. This may explain the difficulty of achieving in whole class teaching the dialogic discourse that is required for effective learning to happen. In fact, research shows such difficulties in classrooms where learners are taught in their home language. How then can whole class teaching achieve the anticipated dialogue in classrooms where learners are taught in a foreign language?

In South Africa, this question has to take into account the prevailing language in education policy debates. South Africa presents a multilingual policy context with monolingual classrooms by choice. By this, I mean that the Language in Education Policy (Government of South Africa, 1997) allows for any of the eleven official languages as recognised by the South African constitution to be used as a language of learning and teaching, LOLT or medium of instruction (MI) beyond the first three years of schooling. Thus, teaching is in the learners' home language until primary year 3 (Grade 3) at which point each school is free to decide on a LOLT according to the local School language policy. The majority of South African schools choose English as the LOLT (Howie et al., 2008). The language education community and policymakers are divided on whether or not Grade 3 is too early for the transition to the English medium of instruction (EMI). Some argue that the persistently poor literacy levels nationally are a consequence of this early transition together with poor teaching of languages generally (see for example, Howie et al., 2008; McKinney & Tyler, 2019; Sibanda, 2017). This situation is not unique to South Africa, most Sub-Saharan African children do not meet the minimum proficiency standards in reading (Trudell, 2016).

Poor language preparation in the lower levels has implications for what is possible in later grades where the teaching and learning of specialist subjects happens in English which is neither the teachers nor the learners home language or language of proficiency. In science, this has implications for the desired learner engagement for meaning-making whether in small group work or whole class teaching (Msimanga & Lelliott, 2014; Probyn, 2016). The challenge to achieve the dialogic discourses alluded to earlier is even bigger in South African science classrooms where learners are not always proficient in the LOLT. How then might science teachers be able to create opportunities for learner talk and engagement in whole class teaching in these multilingual contexts?

In this chapter, I illustrate how two South African teachers attempt to create dialogic discourse in whole class interaction in their multilingual science classrooms. I demonstrate how they leverage the linguistic resources of their classrooms to create opportunities for learner talk. While debates in the context of South African curriculum change tend to pitch teacher centredness (as seen in traditional transmission whole class teaching) and learner-centredness (implied in group work approaches) in tension, with the former viewed as old and undesirable and the latter as new and preferable, I argue that whole class teacher guided approaches have potential to be dialogic rather than transmission so as to achieve meaningful learner talk and engagement. I draw on Mortimer and Scott's (2003) framework for analysing teacher–student interaction in science classrooms to illustrate how the two teachers were able to create such teacher-led dialogic whole class interaction. I also discuss some of the tensions that this raises in the current South African language policy context.

2   Mortimer and Scott's Framework for Analysing Interaction in Science Classrooms

Mortimer and Scott's model categorises teacher–student talk along the dialogic-authoritative and the interactive–non-interactive continuums, recognising four possible teacher communicative approaches during a science lesson; the interactive/dialogic (ID), the non-interactive/dialogic (NID), the interactive/authoritative (IA) and the non-interactive/authoritative (NIA) approaches (Fig. 1).

In the Interactive-Dialogic (ID) communicative approach the teacher engages students in dialogue as s/he explores their ideas; in the Non-Interactive-Dialogic (NID) approach while the teacher is no longer engaging the students interactively s/he continues to review or refer to their ideas elicited during the ID phase; in the Interactive-Authoritative (IA) approach the teacher engages the students usually in a question and answer session, guiding the talk towards a specific scientific view; finally, in the Non-Interactive-Authoritative approach (NIA) the teacher takes an authoritative approach in which only the scientific view is expressed through the voice of the teacher alone, quite akin to the "transmission" mode. According to Mortimer and Scott (2003) dialogic discourse draws learners in, exposes their views and legitimises their talking and thinking—it opens up for genuine learner talk and involvement. Thus, dialogic discourse creates extended interaction which can provide opportunities for learner meaning-making (Scott et al., 2006). The more strictly teacher-controlled authoritative discourse on the other hand is useful in maintaining focus on the scientific story. Successful science teaching must create and draw from both the authoritative and the dialogic discourses (Scott & Mortimer, 2005). This speaks to the tension between the nature of science as an authoritative discourse and the need to engage student ideas as well as create the social interaction (talk) necessary for construction of scientific meaning. Thus, whole class teaching has

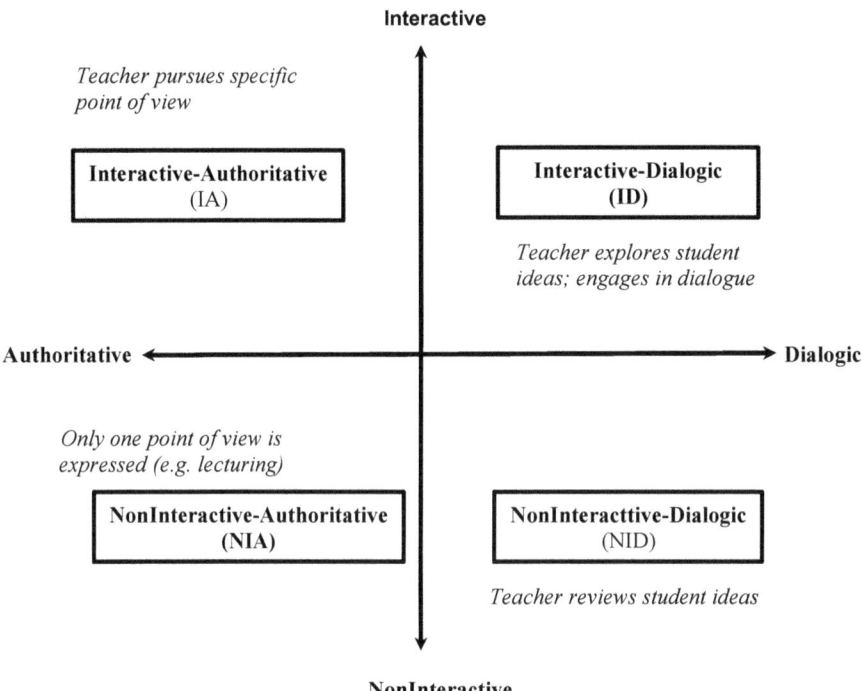

Fig. 1 My visual impression of Mortimer and Scott's (2003) categorisation of classroom talk along the two continuums (Adapted from Msimanga, 2013)

the potential to be both "transmission" (pursuing established science) and dialogic (exploring ideas, understandings and meaning-making).

For each teacher, I characterised the interactive discourse to determine the nature of learner participation resulting from either the IA or the ID teacher communicative approach. In interactive discourse the teacher involves learners in the classroom talk, guiding the discussion usually in a question and answer sequence. In many classrooms, such talk takes Mehan's (1979) traditional IRE (Initiation-response-evaluation) sequences, with mainly chorused, single syllable or yes/no answers. However, in more engaged classroom talk the teacher genuinely engages the learner in sustained IRFRFRF (Initiation-response-feedback) chains in which the F or feedback move speaks directly to learner responses. According to Aguiar et al. (2010) teacher probes (the P move) can open up the interaction closer to a true conversation by encouraging learner questions and through unsolicited learner ideas thus creating IRPRP…E closed chains and the IRPRP open chains. Like the traditional IRE, closed IRPRP…E chains culminate in teacher evaluation of the extended talk, which is typical of IA discourse while ID discourse is characterised by open IRPRPRP chains of genuine uptake of learner ideas without teacher evaluation or judgement. In the excerpts below I show how the two teachers were able to create both closed and open chains in their lessons.

3   The Teachers and Their Practices

3.1   *Mrs. Thoba*

Mrs. Thoba was an experienced Mathematics teacher with 15 years' experience at Grade 11 (16–17 years). She had also been teaching the chemistry component of the Physical Sciences for 5 years to all six Grade 10, 11 and 12 classes in the School. Mrs. Thoba's workload was quite high and while she felt that talking would help her learners engage with the science content and enhance their learning opportunities, she was also anxious about not having sufficient time to cover the content-heavy chemistry syllabus. She was particularly concerned about managing learner talk in a class of 50 learners. She argued that:

> It *(talk)*does provide for effective learning but the only thing is it takes time so if your learners are involved they may talk and talk and it's sometimes difficult to move on with the lesson then you fall behind the time frame to cover all the content in time for exams.

Mrs. Thoba was also concerned about her learners' language abilities. She and her learners were not native speakers of English, the LOLT. They spoke mostly isiZulu and seSotho. Even the teachers hardly spoke to each other in English. However, in class Mrs. Thoba spoke only in English although she allowed her learners to code switch. She said this would help her learners practice English since their final examination are written in English. For her language was a barrier for learner talk in class:

27 Mrs. Thoba: I have observed that outside class they talk a lot maybe it's
                            because they talk in their own language
     Interviewer: Do you sometimes allow them to speak in their own language?
30 Mrs. Thoba: yes I do but then I have to translate for the rest of the class
     Interviewer: You have so many learners with different languages do you
                            understand all the languages?
     Mrs. Thoba: No, like Tsonga I only understand a little. So I sometimes ask
                            another learner to translate

3.2   *Mr. Far*

Mr. Far was an experienced Physical Science teacher with 28 years of science teaching He held a Master of Education degree in Science and Mathematics education. He taught Physical Science Mathematics to Grade 11–12 and was the head of both the Physical Sciences and Computer Technology departments.

Like Mrs. Thoba he taught in a township school and had a big workload but he was keen to use talk strategies in his lessons. He too taught large classes in overcrowded classrooms. For instance, he would have as many as forty-eight (48) Grade 11 learners in a laboratory designed for 25 learners. Most of his learners spoke Afrikaans but the language of teaching and learning in his school was English. Both Mr. Far and his learners did not code switch in class. He spoke often about giving his learners access to a good education so that they could change their personal situations and break out of the poverty that was prevalent in their community. To this end, he incorporated the teaching of values into the teaching of science.

4 Characterisation of Teacher–Learner Interaction in Mrs. Thoba's Lesson

Like in many science classrooms, the main form of learner engagement in Mrs. Thoba's lessons was in response to her questions. However, as she probed and maintained high cognitive demand thinking questions while allowing her learners to use their home languages they began to ask questions and respond to each other's contributions. This was the case in the excerpt below taken from a lesson on bond energy. In this excerpt, the teacher was trying to get the class to resolve a misconception that had arisen from one of the learners, Tahari's answer to the teacher's question. All excerpts are transcribed verbatim in the language in which the utterances were made, written here in italics and an English translation is provided in brackets:

86	T:	When does a po…when does a negative charge form? When does an atom become negatively…?
	Owen:	When two atoms collide?
	T:	Hah?
	Owen:	When two atoms collide
90	T:	And?…
	Owen:	it becomes negative charged…they have one electron
	Tahari:	Eh Maam *manje angithi seziya kholay…seziyahlangana angithi…*(because now they are colliding…coming together)
	Class:	(shuffling and whispering)
	Owen:	That's what we think… *(laughing)*
95	T:	(pointing to Thinta) Let me give you a chance
	Thinta:	Madam I disagree with the statement coz Maam I think when the two (inaudible) the chemical potential energy will increase
	T:	Why…(inaudible)…why do you disagree with the statement?
	Thinta:	It's because Maam…when the…the…the two atoms Maam interact it's impossible for them to be negatively charged
	Tahari:	Maam didn't you say…
100	T:	Why?
	Thinta:	*Azikathintani* (they have not yet touched) Maam
	Class:	Yes…yes

Tahari:	Maam didn't you say when they get closer to each other when they attract the potential energy it will decrease *angithi* Maam?
Thinta:	Its like this… (holding pen and set square apart in each hand)
105 Tahari:	It will decrease…
Thinta:	*Azikathintani* (they have not yet touched) Maam

The excerpt opens with a question and answer session between the teacher and Owen who tries to answer Mrs. Thoba's question about the conditions in which a negative ion is formed. Tahari then joins the conversation and attempts to answer the question providing an explanation which Owen seems to support (turn 94). The teacher ignores both Tahari and Owen's contributions and points at another learner, Thinta to "give you a chance". The subsequent engagement is interesting for this chapter as it illustrates what I saw in a number of Mrs. Thoba's lessons. I refer to this as "ternary interactions" By this I mean that even though learner talk was still channelled through the teacher an interesting form of dialogic discourse emerges which involved the teacher and three learners, Owen, Tahari and Thinta. Instead of the traditional IRE, teacher–learnerX–teacher–learnerY pattern there emerges a teacher–learnerX–learnerY–learnerZ–teacher interaction in the form of an open IRRRR… chain. In this case, the chain spans utterances 97–106, a total of nine utterances where only two were the teacher's. That is, "Teacher-Thinta-Tahari-Teacher-Thinta(Class)-Tahari-Thinta-Tahari-Thinta". I view this as typical dialogic discourse between the teacher and not one but three learners. It is also significant that this dialogue plays out in two languages.

Earlier in the same lesson, the teacher had indicated to Tahari that it was alright for her to speak in her language. The learner's response is of interest to the argument being made in this chapter. The learner expressed reluctance to use her language. Unfortunately, she did not complete her sentence and there was no time to interview her after the lesson to understand from her the difficulty that she had. She was responding to a question, "What is the net force between the two atoms?" (the teacher had put a diagram on the board of two atoms with a line showing the distance between them):

53 Tahari:	Because of there is no attraction…they are not…they are far distant…so they…they are…I can't explain Ma'am…
Teacher:	say it in your language its fine
55 Tahari:	In my language Ma'am …? Ma'am in my language its so…
Class:	(Learners laugh)
Tahari:	ok… fine Ma'am let me say it in my language…ok fine *Ma'am tinekule Ma'am atihlangananga ti* (inaudible) Ma'am but Ma'am *loko tingahlangananga tahari* constant *tahari* net force *yatona tahari zero* and so *loko tita atrakthana loko setita hlangana tiya atrakhta ke yikhona tingataba…tingataba…*
Teacher:	(Teacher finishes in English) The forces will then attract …

Tahari: Yes

Teacher: (Teacher translates) Ok what she means is…if they are apart then their potential energy is zero because there are no forces acting between the…two atoms

The excerpt opens with Tahari trying and failing to explain her understanding of the diagram in English. In turn, 54 the teacher then "grants permission" to "say it in your language its fine" to which the learner responds "In my language Ma'am …? Ma'am in my language it's so…".

The teacher's act of encouraging Tahari to give the explanation in her language signals that it is acceptable in her class for science to be discussed in a language other than English. In the context of an English-only school policy, the teacher's act was non-compliant and the learner's reluctance could also signal that she did not view this as appropriate practice in a science classroom. However, when she eventually engaged in her language she was able to explain her understanding of the concept and provide an acceptable answer to the question.

It would appear that Mrs. Thoba's open ended questioning techniques together with her openness to learner language use in her classroom created a conducive environment for both the ternary interactions in which learners shared the social space freely and the dialogic engagement that played out in this whole class discussion. Such a sophisticated approach to whole class discussion is not easy for teachers to enact and sustain. Research must document evidence of such practice and the conditions that support it. In turn, teacher education programmes must find ways to articulate and make available such approaches for inclusion in the new teachers' pedagogical toolkit.

In the next section, I illustrate how the other teacher, Mr. Far worked differently to achieve similar dialogic discourse in his multilingual classrooms.

5 Characterisation of Teacher–learner Interaction in Mr. Far's Lessons

Mr. Far questioned more than Mrs. Thoba and used more open questions often persistently probing the learner until he got a response. He often took learners' ideas and understandings seriously—although he ignored some learners' contributions he generally valued learners' ideas and used them to direct the course of the lesson. His interventions that produced the different teachers' communication styles were a mix of elaborative and evaluative teacher responses to learner contributions. Mr. Far responded more elaborately to learners' contributions most of the time. He encouraged learners to evaluate and critique their peers' ideas, often foregrounding learners' ideas for interrogation by their peers. This created the potential for increased student participation as learners gained confidence in evaluating each other's ideas. However, Mr. Far asked the majority of the questions himself soliciting some learner questions along the way.

The lesson illustrated in this chapter was an introduction to momentum and although teaching was in a normal classroom, Mr. Far had the learners conduct simulated collisions. This was typical of Mr. Far's practice. He invariably involved his learners in practical work, a practice that was not common in other township schools in Mr. Far's context. Mr. Far also did something else for his learners. He would often say "Why am I doing this again? Because I'm gonna come back to the molecule later …" and he always did. Thus he provided clear links for his learners between the different parts of the lesson and the different concepts being considered in the lesson. While this practice may not be specific to teachers of ESLs it becomes even more important for making the connections clear for learners who experience challenges with the language of instruction.

The excerpt below illustrates these characteristics of Mr. Far's practice. The lesson started with a session to "… just refresh quickly":

3   Teacher: according to eh the definition of momentum it can be regarded as a measure of the product of the mass and the velocity. Now Kelvin if you think about mass and velocity think about mass in terms of the quantity can we regard mass as a vector quantity or is it a scalar quantity?

      Class: *(learners shouting)* vector … scalar… scalar … vector

5   Teacher: now I will say that again think about it carefully

      Class: *(talking among themselves)*

      Teacher: think about mass where do we find mass because she has used the words mass and velocity

      Len: scalar Sir

      Teacher: why?

10 Kelvin: because yah the mass is got size

      Ben: yah its …

      Martin: mass is got size

      Teacher: so why am I asking this? Because our biggest problem that we encounter is that most of us cannot distinguish between this and that *(pointing to the words scalar and vector)*. So let's just refresh quickly. Len you said this is scalar why are you saying its scalar?

      Sisa: because it has size only and no direction …

15 Teacher: thank you very much so we only have size which is also …

      Busi: magnitude

      Teacher: magnitude. So here we have size or magnitude

      Busi: no direction

      Teacher: so this is what we are having we are having a scalar quantity which is mass and velocity a vector. Now remember what is momentum in real sense? We wanna make it simple in our heads coz we are labelling this thing. What can you describe momentum as? A simple word?

The teaching purposes in this episode were mainly to introduce the topic of the day, starting with a review of prior knowledge of key concepts like scalar and vector quantities. The teacher's predominant interventions were elicitation and evaluation with mostly closed and some open questions in an IRE/F discourse with the occasional IRPRPE closed chain. For example, in the first IRE triad in turns 3–5 the teacher opened the discussion with a closed question to Kelvin as to whether mass was a vector or scalar quantity (I), to which he got a mixed reaction from the class in general, with some learners shouting "vector" and others shouting "scalar" (R). The teacher's evaluative feedback (E) in turn 5, "I will say that again think about it carefully" was seemingly interpreted thus by the class who then engaged in private discussions among themselves to "think about it carefully". The teacher followed this with a clue about mass, which served as an initial move for an IRPRRRE chains. I see later you do refer to "probe" (p. 8, para 2) but perhaps clarify when you first use it.) discourse between Mr. Far and five learners, Len, Kelvin, Ben, Martin and Sisa in turns 7–14. The chain stopped when the teacher made an evaluative statement in turn 15 thanking (and affirming) Sisa for her answer, thus indicating the end of the discussion. The evaluation was followed by a summary in turn 19 and a new initiate move for a discussion to find a simple word to describe momentum.

The next 15 min of the lesson were spent with learners simulating collisions with various objects and the teacher talking them through their observations. The following short excerpt from the practical activity illustrates again how Mr. Far had the learners not only make their own observations but they had to explain and write down their observations:

67	Teacher:	Now this is what I want you to do. Take out anything you have in your pocket. Either you have two pens in your pocket take it out you have two coins whatever you have take it out put it in front of you. This is the task you need to have those two objects that you have in front of you make a collision make a collision then if you do that you have to look what type of collision you have whatever you have in front of you. So I will just walk around and see if you are with me. So put your objects two of them and then you collide those objects and look at the type of collision. *(walks around)*
	Nikitha:	*(inaudible)*
	Teacher:	Nikitha Nikitha is asking why do we need two?
	Nikitha:	*(inaudible)*
	Teacher:	yes just throw it just let it collide Nikita
	Teacher:	money money throw your money and remember always what do we need to do? We need to *write down isn't it so*? what we are seeing or what we are observing then we work from there. Right? its a small experiment

Mr. Far created what he called "a small experiment", asking learners to throw whatever they had and observe the type of collision. He walked around and instructed learners to write down what they saw (Turn 72). The nature of the practical activity in itself was conducive to learner participation as each learner had to conduct his/her own "small experiment" and observe. Also the fact that the teacher walked around as he talked the learners through the activity ensured that all participated. This was a highly interactive and dialogic lesson. Learners together with the teacher engaged in exploring learners' ideas about collisions and together negotiated understandings of the terms as they talked about each of them in turn.

In the next two excerpts, Mr. Far's teaching purpose shifted from exploring learners' ideas and allowing them to explore their own ideas to develop the scientific story. His communicative approach changed from fully interactive-dialogic to alternating between ID and IA communicative approaches:

77 Walter: Sir my observation Sir I had a two rand in one hand Sir
    Teacher: yes different objects
    Walter: so when I collide them the one rand went away which means the
               two rand is heavier than the one rand
80 Teacher: now describe to me exactly what you mean going away
    Walter: the two rand pushes the one rand away Sir
    Teacher: so someone else .... (inaudible)
    Alan: equal masses I had two pens
    Teacher: you had two pens of equal masses so we have one scenario
               different masses then we have the second scenario with same
               Masses
85 Alan: ... *(inaudible)* different direction
    Teacher: so you had *(inaudible)* this way and then it went different
               direction. Any other person? Yes P?
    P: ... same as this
    Teacher: how can it be same like this? How is it possible
    P: *(inaudible)*
90 Teacher: ok
    P: and they had the same mass ...*(inaudible)*
    Teacher: right *(in raised voice)* he is saying he's doing this and I actually
               like this he says this might be his two pens throw them together
               they collide and they went opposite direction. Do you all see this?
    Class: yes
    Teacher: now describe this type of collision this type of collision *(teacher
               hits his fists together and moves them in opposite directions)*

95 Class: *(inaudible)*
    Teacher: this type of collision
    Altus: elastic
    Teacher: now give me your definition of elastic
    Altus: of elastic Sir?

In the episode above the learners were now reporting back on their observations of simulated collisions. The excerpt started with the teacher checking that the learners had finished writing their observations and then Walter describing what he had done and seen when he made a one-rand coin and a two rand coin collide. The teacher adopted a mix of IA and ID communicative approaches, eliciting learner ideas, questioning, probing and evaluating some but accepting others without evaluation. An IRFRPR open chain discourse ensued between him and Walter in turns 77–82. This open chain discourse resulted from an ID communicative approach commenced with an initiate move by the teacher (I) asking Walter to give a report, to which Walter responded in turn 77 (R) with a description of the coins that he had used, interrupted by the teacher in turn 78 with elaborative feedback (F) "Yes different objects". In that statement, the teacher affirmed Walter's report with "yes" and then elaborated on it pointing out the fact that the objects were different. This would serve to mark the idea suggesting to Walter and the rest of the class that the difference was significant and would shift the talk from dialogic to authoritative creating the tension that Mortimer and Scott (2003) argue must exist if the teacher has to explore learners' ideas while pursuing the scientific story. In this case the teacher was indeed exploring learners' ideas about their own collision but also pursuing the teaching purpose of developing the scientific story on the basis of those ideas.

When Walter explained that his one-rand coin "went away" because the two rand coin was heavier (response, R) the teacher probed (P) for an explanation of "going away" and then accepted the explanation (R) without evaluating it, moving on to solicit other learners' reports (Turn 82). Alan's response in turn 83 confirmed that he had noted and taken up the teacher's point about the fact that Walter's coins were different. He started his report, "Equal masses I had two pens" and the teacher communicated his agreement by revoicing Alan's opening statement, elaborating on it, "we have one scenario different masses then we have the second scenario with same masses" (Turn 84). The teacher again took up and elaborated on Alan's next point that his pens went in different directions after the collision, again marking and foregrounding the idea. Finally, in response to a third learner, P who is giving his report gestured with his hands to illustrate the movement of the objects, the teacher raised his voice and called the attention of the class to P's gestures.

In turn 92, the teacher made several interventions that finally linked his two teaching purposes, to elicit learner ideas and to develop the scientific story. He affirmed P, "Right *(in raised voice)* he is saying he's doing this and I actually like this …", then he repeated P's gestures while paraphrasing P's contribution, marking the idea as important, "and then he says this might be his two pens throw them together they collide and they went opposite direction. Do you all see this?" The teacher then took up P's idea (the gesture) and used it to get the class to think through and name the collision. Finally, Altus gave the correct scientific name for that type of collision as "elastic" (Turn 97). The teacher's next turn inevitably opens up a new episode to define an elastic collision. This kind of interaction continued throughout the lesson as the class identified the different types of collisions, the different energy changes and finally defined momentum itself. To end the lesson the teacher engaged the class in a non-interactive session taking an NIA communicative approach to pull together the different concepts covered and to get them to start thinking about the forces involved in the collisions.

Mr. Far took a dialogic approach to encourage learner participation and thinking by involving them in practical activities. He would then switch to an authoritative style to develop the scientific story and explain new terms to the learners. His interventions tended to be evaluative resulting in mostly IRF triads and some closed IRPRP...E chains typical of authoritative communication. Mr. Far always made the connections clear; showing how the concepts were linked; showing links between and within the lesson to illustrate continuity as well as providing affirmations to promote learner emotional engagement. He always had his learners write and he often engaged in meta-talk. Mr. Far also made the most "small talk" with his learners. He created the kind of classroom environment described by Bishop and Denley (2007) where science learning was fun, and both teacher and learners dared to do things differently. In my view, he was able to open up classroom interaction for non-English learners to experience the kind of learner-centred classroom anticipated by the curriculum. Although he and his learners did not code switch or draw on their common language, Afrikaans, he managed to engage with language in ways that created opportunities for non-English learners to learn science.

6   Discussion and Conclusions

Teaching science or indeed any subject in a language that learners are not proficient in is difficult even for the best of teachers. However, having to do so in the constrained teaching and learning environments that prevail in many low socio-economic contexts is an even bigger challenge. In the South African context, the situation is compounded by the many historic factors including a long tradition of much teacher talk and no learner talk, predominant transmission methods and teaching to the test. The result is poor achievement in science as learners simply memorise and regurgitate concepts in examinations. Thus, small group work has been advocated by many to create opportunities for learners to engage and make sense of the science content for themselves. However, the uptake of group work has remained low due to large classes, overcrowding, teacher anxiety about insufficient time to cover the curriculum and the challenge of working in multilingual classrooms. Hence the persistence of whole class teaching. This chapter has provided evidence that whole class teaching has potential to meet the objectives of learner involvement, learner-centredness as espoused in the South African curriculum. This data illustrates what I see as pockets of success with whole class teaching and how teachers can and do leverage learner languages to involve learners in the discourse of the classroom.

Mortimer and Scott's framework enabled a nuanced understanding of the nature of learner engagement when teachers open up the classroom talk for genuine learner interaction. The framework distinguishes between the two teachers' practices and how they worked in ways that were similar in some respects but different in others. For instance, both Mr. Far and Mrs. Thoba were able to engage the learners in co-constructing the scientific story through whole class dialogic talk as they directly responded to their contributions and wove these into the scientific explanation. An important difference was how Mrs. Thoba encouraged learners to express ideas in their home language while Mr. Far engaged his learners in multimodal activities including writing and practical activities to enable meaning-making in an unfamiliar language. While the data illustrates how these teachers, particularly Mrs. Thoba worked with learner languages not as a barrier but as a resource in learning science, her practices are not unproblematic in the current language policy context in South Africa. While the multilingual provisions of the Language in Education Policy provide impetus for Mrs. Thoba's practices, the realities of the monolingual policy context at the local school level render such approaches "illegal" and hence Tahari's question on the appropriateness and/or efficacy of her home language in a science discussion. Yet, research evidence abounds on the value of learners' languages at least for engaging with difficult concepts in science classrooms.

In the current policy context in South Africa teachers often find themselves having to choose between what they know about the pedagogical benefits of using their learners' languages and the school policy requirements on language use in the classroom. Thus, teacher efforts as illustrated in this chapter remain uncelebrated, poorly documented and not available to others especially to beginner teachers to adopt as part of their toolkit. In other words, current policy requirements stifle teacher agency towards achieving the very learner-centred methods espoused by the curriculum. Current policy debates on language in education and on language teaching in general must include how to enable context-informed choices on language use in the teaching and learning of specialist subjects like science. Future research must explore the nature of support required by teachers who do open up their classrooms for genuine multilingual engagement. Likewise, teacher education programmes must prepare teachers to manage classroom interaction in ways that create the desired dialogic discourse. More importantly both research and teacher education must address both teachers and policymakers concerns about the perceived repercussions of using learners' home language in teaching and learning on learner success in national exit assessments which are administered in English for the majority of South African learners.

Acknowledgements I am indebted to both Dr. Lizzi Okpevba Milligan and Dr. Margie Probyn for a critical review of an early version of this manuscript.

References

Aguiar, O. G., Mortimer, E. F., & Scott, P. (2010). Learning from and responding to students' questions: The authoritative and dialogic tension. *Journal of Research in Science Teaching, 47*(2), 174–193.

Bishop, K., & Denley, P. (2007). *Learning science teaching: Developing a professional knowledge base.* UK: McGraw-Hill Education.

Brodie, K. (2000). Teacher intervention in small-group work. *For the Learning of Mathematics, 20*(1), 9–16.

Department of Education. (2012). *The National Curriculum Statement (NCR): Curriculum and assessment policy statement grades 10–12 physical sciences.* Pretoria: Department of Education.

Department of Education. (1997). *Language in Education Policy 14 July 1997.* Pretoria: Government of South Africa.

Howie, S., Venter, E., & Van Staden, S. (2008). The effect of multilingual policies on performance and progression in reading literacy in South African primary schools. *Educational Research and Evaluation, 14*(6), 551–560.

Lyle, S. (2008). Dialogic teaching: Discussing theoretical contexts and reviewing evidence from classroom practice. *Language and Education, 22*(3), 222–240.

McKinney, C., & Tyler, R. (2019). Disinventing and reconstituting language for learning in school science. *Language and Education, 33*(2), 141–158.

Mehan, H. (1979). 'What time is it, Denise?': Asking known information questions in classroom discourse. *Theory into Practice, 18*(4), 285–294.

Milligan, L. O., & Tikly, L. (2016). English as a medium of instruction in postcolonial contexts: Moving the debate forward. *Comparative Education, 52*(3), 277–280.

Mortimer, E., & Scott, P. (2003). *Meaning making in secondary science classrooms.* Berkshire, UK: McGraw-Hill Education.

Msimanga, A. (2013). *Talking science in South African high schools: Case studies of Grade 10–12 classes in Soweto* (Doctoral dissertation, University of the Witwatersrand, Faculty of Humanities, School of Education).

Msimanga, A., & Lelliott, A. (2014). Talking science in multilingual contexts in South Africa: Possibilities and challenges for engagement in learners home languages in high school classrooms. *International Journal of Science Education, 36*(7), 1159–1183.

Probyn, M. J. (2016). *Language and the opportunity to learn science in bilingual classrooms in the Eastern Cape, South Africa* (Doctoral dissertation, University of Cape Town).

Scott, P., & Mortimer, E. (2005). Meaning making in high school science classrooms: A framework for analysing meaning making interactions. *In Research and the quality of science education* (pp. 395–406). Springer, Dordrecht.

Scott, P. H., Mortimer, E. F., & Aguiar, O. G. (2006). The tension between authoritative and dialogic discourse: A fundamental characteristic of meaning making interactions in high school science lessons. *Science Education, 90*(4), 605–631.

Sibanda, J. (2017). Language at the grade three and four interface: The theory-policy-practice nexus. *South African Journal of Education, 37*(2), 1–9.

Trudell, B. (2016). Language choice and education quality in Eastern and Southern Africa: A review. *Comparative Education, 52*(3), 281–293.

Audrey Msimanga is Associate Professor of Science Education, currently the Head of Education at Sol Plaatje University as well as a Visiting Researcher at the University of the Witwatersrand. Audrey has worked in Biology research and then in Science Education for over 30 years. She is the lead researcher on the "Talking Science project" which Audrey's research seeks to understand the role of social interaction in science learning; the potential for classroom talk to mediate learner meaning-making as well and the role of language in science teaching and learning in multilingual classrooms. Audrey is currently an Associate Editor for the Journal for Research in Science Teaching (JRST).

Multiple Monolingualism Versus Multilingualism? Early Grade Mathematics Teachers' and Students' Language Use in Multilingual Classes in South Africa

Ingrid Sapire and Anthony A. Essien

Abstract The language of learning and teaching (LoLT) in schools, and more specifically, the language used in the teaching of mathematics, is the focus of academic and public debate because language is political. Constrained by the Curriculum and Assessment Policy Statement's (CAPS) interpretation of the Language in Education Policy (LiEP), the South African school system can be seen as a system of multiple monolingualism rather than a truly multilingual system. It can be argued that this is the general trend in developing countries previously controlled by colonial powers. In South Africa, there has been a dearth of research studies undertaken in early grade mathematics classes to investigate language use in these classes and whether it aligns with the languages spoken by the teachers and students present in those classes. The research reported in this chapter investigated language use in early grade mathematics classes. Data collection was in three parts and involved a survey of the language background and perceptions about language use in mathematics teaching, worked solutions of number patterns questions, and a translation activity on number patterns. This was carried out in Grade 3 and Grade 4 classes (two of each) in a sample of 20 schools in three districts in a Province in South Africa (with LoLT IsiZulu, Setswana and English). Altogether 62 teachers and 2891 students (with 13 student home languages and 7 teacher home languages) participated in the study. Through our findings, we shed light on ways in which the interpretation of the curriculum language policy (which embodies a monoglossic language ideology) influences language use in South African schools and to what extent this policy determination in its current form promotes the multilingual education system envisaged by the policy.

Keywords Multilingual mathematics teaching · Monolingualism · Multilingualism · Translanguaging · Language policy

I. Sapire (✉) · A. A. Essien
Wits School of Education, University of the Witwatersrand,
Johannesburg, South Africa
e-mail: ingrid.sapire@wits.ac.za

A. A. Essien
e-mail: anthony.essien@wits.ac.za

© The Author(s), under exclusive license to Springer Nature Switzerland AG 2021
A. A. Essien and A. Msimanga (eds.), *Multilingual Education Yearbook 2021*, Multilingual Education Yearbook,
https://doi.org/10.1007/978-3-030-72009-4_5

1   Introduction

The language of learning and teaching (LoLT) in general, and more specifically, the language used in the teaching of mathematics, has been the focus of academic and public debate in different parts of the world due to the political nature of language (Clarkson, 2016; Kajoro, 2016; Phakeng, 2017). Providing a succinct overview of the issues for mathematics teaching and learning within the changing policy land-scape of Papua New Guinea, Clarkson (2016), for example, shows that mathematics teaching in this context has never been divorced from political decisions made by others outside of the education system. Studies have also shown that the implementation of some of the language policies in mathematics classrooms has been fraught with difficulties (Essien, 2018). In South Africa, the Language in Education Policy (LiEP) was developed early in the new democratic South Africa and was designed to promote multilingualism in schools so that all South African students can be taught in their home language (Department of Education, 1997). However, implementation of the policy has not always achieved this goal since this implementation occurs through the choice of a LoLT by schools, which does not always coincide with the home language of all students at the school. It is not surprising that teachers still say things like *English is used as the main language in Maths because some words cannot be translated to IsiZulu.* What comes to mind here is the point made by Bamgbose (1999, p. 18) who argues that the major problem with language in Education policies in Africa is the tendency to equate planning to 'policy making alone, while implementation tends to be treated with lack of serious concern or even downright levity'.

A report on the status of the LoLT in schools was published in 2010 by the Department of Basic Education (DBE) of South Africa, based on Education Management Information System data gathered over the period 1998–2007. This report showed that in 2007, 20% of students in South Africa who use an indigenous language as the LoLT were not yet being taught in their home language (DBE, 2010). This figure stood at 18% in the updated report on the status of the LoLT in South African schools in 2016 (Sapire & Roberts, 2017). Language use in the early years of schooling in South Africa is thus constrained by the CAPS interpretation of the Language in Education Policy that has resulted in a system of multiple monolingualism (Sapire & Roberts, 2017). As Heugh (2014) notes, this is the general trend in developing countries previously controlled by colonial powers. What is meant by multiple mono-lingualism is a system where many languages are used as the LoLT but only one can be used at a time. In other words, multiple monolingualism entails that even though a person may be proficient in multiple languages, only one of these languages should be used at any one time, as if the speaker is monolingual. In the school system, it entails that one class must be taught in one language and mixing of languages is not encouraged. This is contrary to the literature on multilingual mathematics education which shows that the use of translanguaging practices, which involve drawing on the full language repertoire of teachers and students (using more than one language as a resource when speaking or writing), enables more effective learning in multilingual classes (García & Wei, 2014; Phakeng, 2017).

In South Africa, although there has been ongoing reporting on the status of the LoLT in schools, there has been a dearth of research studies undertaken in early grade (Grades R to 3) mathematics classes to investigate the nature of language use in these classes and whether it aligns with the languages spoken by the teachers and students. The larger study on which this chapter draws (based on the first author's current doctoral research) investigates the language use of Early grade mathematics teachers and students and how this language use aligns with the language in education policy. By *language use* we mean the words chosen by speakers to express themselves. This encapsulates the full range of words used when speaking or writing, including things such as the use of multiple spoken or written languages and formal or informal expressions when speaking or writing mathematically. The purpose of the present chapter is specifically to investigate language use of Grades 3 and 4 teachers and students on the topic of patterns evidenced in a dataset which included a survey of the language background information and perceptions about language use in mathematics teaching, a series of three number patterns questions, and a translation activity based on core mathematical terminology linked to the topic of patterns (number and geometric). In order to do this, we were guided by the following research questions:

- What is the relationship between the LoLT and home language of the teachers and students?
- What variation in language use is evident when teachers and students solve mathematics number pattern problems and provide explanations?
- What variation in language use is evident when teachers and students translate selected mathematical terminology (words and phrases) on the mathematical topic of patterns?

2   Theoretical Orientation

Language ideology was used as a framework for the study. Language ideologies vary and they are linked to the speaker's (or writer's) orientation towards language. García and Wei (2014) have linked monoglossia (and on the other side of the spectrum, heteroglossia) to ideology. They do this by speaking about the ways in which language is treated. Monoglossic ideologies treat languages as bounded autonomous systems without regard for the actual language use of speakers, while heteroglossic ideologies respect multiple language use practices in interrelationships (García & Wei, 2014). A monoglossic ideology is linked to a purist view of language which holds that one pure language can be used to express oneself meaningfully. A heteroglossic ideology is one linked to a pluralist view of language which holds that speakers who have a language repertoire consisting of more than one language resource are able to (and do) draw on multiple languages when they speak. A heteroglossic ideology acknowledges linguistic diversity.

Language use practices can be understood in relation to language ideologies whether the users of language are aware of it or not, because language ideologies may be articulated or embodied (Kroskrity, 2004). Articulated ideologies are evident when speakers are questioned about their language use choices, and they give explicit reasons for these choices and indicate that they are aware of their language use choices. The explanations they give express their articulated ideology with regard to language use. For example, a teacher who is questioned about the way in which she has given a particular explanation during a lesson who is able to give reasons for why she used code-switching is expressing an articulated ideology. Embodied ideologies are evidenced when speakers use language in a particular way. Such language use may be conscious or unconscious. This is possible since speakers may give conscious thought to choices they make in relation to their language use or they may just speak without consciously thinking about the language use choices they make while speaking. An example of a consciously embodied ideology is found when a teacher explains a concept to a multilingual class and consciously chooses to give a particular definition in both the home language of the majority of the students as well as in English. An example of an unconscious embodied ideology is found when a teacher, drawing on her full language repertoire (using translanguaging), explains a concept to her class. Embodied ideologies can also be identified in written material—evidenced in the language use in a written artefact. Simply put, language use may be multilingual or monolingual. Multilingual language use may take on many forms (primarily seen as code-switching and/or translanguaging) but monolingual language use takes on one form—the pure use of only one recognised language. In this chapter, we report on *language use* choices made by teachers and students in the sample schools. We use the term *mixed language use* to refer to situations where code-switching and/or translanguaging practices are used.

3   Bilingualism/Multilingualism in the Learning and Teaching of Mathematics

A growing number of recent research studies have positioned multilingualism as an advantage rather than a problem/deficit. It is important to note that this has not always been the case. Research in the early 70s (and before that) on the effects of bilingualism in learning and teaching postulated that bilingualism has a negative cognitive effect (Lyon, 1996). The assumption made by the researchers at the time was that the human brain was compartmentalised and as such, the more language a person is exposed to, the more compartments the brain is divided into and hence, the lesser the cognitive, educational and even linguistic development. Such research (or theory—for example, Macnamara's 'balance effect' theory) tended to conclude (or postulate) that the negative cognitive effect of bilingualism was due to the dissipation of the stock of their available intellect in knowing two languages or in learning an additional language (Cummins, 1979). Up until the late 70s, Macnamara (1977) for example, held that thought and language were distinct because the former is abstract with respect to the latter. The implication of this, notes Macnamara (1977), is that a child must 'develop somehow both the domain of thought and, separately, the domain of language' (p. 2). It could be argued that one of the most significant works in the domain of bilingualism and cognition was carried out by Cummins (1979) in his postulation of the threshold theory. The threshold theory attempted to explain why some studies reported bilingual students as cognitively disadvantaged compared to monolingual students while others reported bilingual students as more cognitively advantaged than their monolingual counterparts.

A growing body of research on multilingual education has continued to move away from deficit theories of multilingualism to address issues on how the linguistic resources multilingual students bring to class can be adequately harnessed to provide high-quality mathematics education to such students. In other words, recently, more and more, research now focuses on multilingualism as a resource in multilingual mathematics classrooms (Barwell, 2018; Planas, 2018; Prediger et al., 2019; Ryan & Parra, 2019). This research is pertinent to this chapter since the schools where the research was carried out were richly multilingual. Most of these researchers, however, are quick to highlight the complexity of learning and teaching in multilingual class-rooms where students are still learning English as a language and simultaneously learning mathematics both as a discipline of knowledge and as a language.

Students who come from homes where the LoLT is the only language spoken at home are familiar with the linguistic structures they encounter in the mathematics classroom (Barwell, 2018). Research (Barwell et al., 2007; Robertson & Graven, 2018) has shown that this is not the case with students whose home language is not the LoLT. Where the home language is not the elected LoLT, students must deal with the additional constraint of not being fluent in the LoLT—unlike students whose home language is the LoLT, and who are already familiar with the language as well as its linguistic structure. The complexity of the match between home language and LoLT in South Africa is complex since there are 11 official languages and many people speak at least four different languages. In early grade classrooms, such as those in the schools that participated in the survey, the schools may have elected an indigenous language as LoLT but the student population may not all share a common indigenous language. Students may be familiar with the LoLT but it is not their home language in that they may speak an indigenous language that is not the official LoLT of the school. Another issue of contention is that they may speak a language dialect that has not been officially recognised for use as a LoLT (Mojela, 2008; Lafon & Webb, 2010). Mojela writes about the dialects that are associated with the Sepedi language (also known as Sesotho sa Leboa) and argues that 'the Sesotho sa Leboa standard language is more of a second language than a mother tongue to these communities' (2008, p. 125). All students have to deal with the structure of the mathematical language (Pimm, 1981) but additional language students have to deal not only with learning the mathematical concepts, but also the language in which these concepts are embedded (Barwell et al., 2007).

Whether a student will have the opportunity to learn mathematics in their home language is determined by the language policy that guides (and controls) language use in schools. The South African LiEP intended to allow an opening up of the LoLT and a move towards multilingualism in schools that reflected the multilingual South African society and as such embodies a heteroglossic language ideology. Trudell and Piper state that, 'language policy development is actually just one of several intentional activities that have a bearing on the use and development of languages' (2013, p. 2). They also argue that 'in the school context, language ideology plays out in the choices made by teachers, education authorities and parents regarding the language(s) they want to see used in the classroom' (p. 1). Language policy is determined by political players (Mojela, 2008) and it is driven among other things by ideology. In this chapter, we investigate language use in order to gain insight into the way in which the CAPS' interpretation of the LiEP (which embodies a monoglossic language ideology) controls language use in South African schools and whether or not this policy determination in its current form promotes the multilingual education system envisaged by the policy.

4   Methodology

The research design was that of a descriptive study based on a three-part data collection method involving (1) a survey of the language background information and perceptions about language use in mathematics teaching; (2) a series of three number patterns questions and (3) a translation activity based on core mathematical terminology linked to the topic of patterns. The study is descriptive as it gives insight into language use although it does not answer questions about the effects of particular language use as there was no pre- or post-test component to this study. Cai et al. (2009) state that, '[o]ne of the most robust findings of education research is that students learn best that which they have the opportunity to learn' (p. 233). Knowledge of language creates the opportunity to learn. This study, by investigating language use, gives insight into the language used in (and necessary for) the learning and teaching of mathematics. The curriculum topic of patterns was selected to contain the scope of the study. Low performance on pattern items in some national standardised tests in South Africa was noted (e.g. DBE, 2014) hence there was a perceived value in investigating language use in relation to this topic. Data were collected in a sample of 20 schools in three districts in a province in South Africa (with LoLTs IsiZulu, Setswana and English) in Grade 3 and Grade 4 classes (two of each, hence 80 classes participated altogether). IsiZulu and Setswana were chosen to represent the two main indigenous language groups in South Africa. Altogether 62 teachers and 2891 students (aged between 7 and 9) participated in the survey which was carried out in 2017. Ethical clearance was granted for the study and all of the necessary consents were obtained.

The data collection tool was a questionnaire. The *first part* of the questionnaire collected information on the LoLTs of the schools and the language background of the participants. The teachers/students were asked to name their main spoken and additional spoken languages. Teachers also had to note the different languages they use/have used to teach mathematics and to expand on and explain this use of language. This biographical information was gathered to get accurate information to provide the backdrop to the language ideological enquiry.

The *second part* of the questionnaire consisted of a set of three questions about number patterns which yielded data on the use of mathematical register in the form of answers and mathematical explanations, since the questionnaire called for explanations of answers. The primary goal of the questionnaire was the collection of language use data—it was not designed to 'test' mathematical knowledge. The level of the questions was thus specifically chosen so that teachers and students should be able to answer them easily. The mathematical questions were presented using a bilingual format (IsiZulu/English and Setswana/English) and participants were told that they could use the language of their choice to complete the questionnaire. This suited the aims of the research as it allowed freedom of language use, leaving the choice in the control of the individual completing the questionnaire. Written responses yielded data on the active use of the language of mathematics. The specification of the mathematics questions is briefly discussed next since it provides the context for analytical interpretation of the responses given by the teachers and students (Fig. 1).

The first two questions present sequences that need to be completed. In question 1, the sequence is made of 3-digit numbers (Grade 3 number range), but the sequence grows by an increase of 2 between successive terms, which is easy to identify. This presents low-level cognitive activity. Question 2 presents a sequence of 2-digit numbers (lower cognitive demand) and the instruction is given to 'extend' the pattern. The rule for the sequence is easy to identify (increasing in 3s) but the answer involves bridging ten into the 3-digit numbers (from 99) which calls on more highly developed number sense and raises the cognitive level of the question. The third question calls for the identification of a number which 'does not belong' to a given sequence. The number range (2-digit numbers) and rule for the sequence (increase by 3) are not complex but the identification of a number that 'does not fit' in a given sequence of numbers is more complex as it involves recognition of a rule and reasoning using the rule to exclude a particular term in the sequence of numbers. This raises the cognitive complexity of the question as alternative rules may be mistakenly identified (such as odd/even numbers) and applied.

The *third part* of the data collection was a translation activity which included a list of 20 words that half of the participants translated from IsiZulu/Setswana to English and the other half translated in the opposite direction. There were thus four different instruments (two language pairs and two directions of translation). The chosen words ranged in difficulty and familiarity. This part of the questionnaire suited the aims of the research because it provided for open-ended translations in both directions between languages and hence it yielded data showing variations in language use.

Questions 1 and 2 from the IsiZulu version of the questionnaire.

1. Fill in the missing number./ Faka inombolo engekho:
    228, 230, 232, _____ , 236

    Explain how you got your answer:
    Chaza ukuthi uyithole kanjani impendulo yakho

    _____

    _____

2. Extend the pattern./ Khulisa leli phethini
    93, 96, 99, _____ , _____ , _____

    Explain how you got your answer:
    Chaza ukuthi uyithole kanjani impendulo yakho

    _____

    _____

Question 3 from the Setswana version of the questionnaire.

3. Underline the number that does not belong to the pattern.
    Thalela dipalo tse di sa weleng mo dipateroneng.
    3, 6, 11, 15, 18

    Explain how you got your answer:
    Tlhalosa go re o fitlhetse karabo ya gago jang:

    _____

    _____

Fig. 1 Exemplar bilingual number patterns questions: Setswana and IsiZulu

5   How Data Was Collected, Captured and Analysed

The research instruments were sorted and prepared for digital capture by a multi-lingual research team to facilitate analysis. Digital data were cleaned by the project team leader assisted by the research team in preparation for statistical analysis. Since quantitative and qualitative data were collected, two forms of analyses were carried out.

A descriptive analysis was undertaken on data from the *first part* of the questionnaire on LoLT, HL, other spoken languages and teaching languages in order to draw up tables and graphs that could be used to give descriptive contextual findings. Responses to the open-ended questions on teacher language use were all typed up verbatim for qualitative analysis. These responses provided a qualitative dataset relating to teachers' language use preferences that was analysed using thematic content analysis, through which common themes emerging from the dataset were written up. The common themes, such as 'I use IsiZulu/Setswana/English to teach because it is the LoLT of the school'; 'I use mixed languages because it helps my students to understand better' and so on, were found by systematically combing through all of the responses and categorising them according to the sentiment they expressed. Themes emerging from the qualitative teacher perception data were used to comment on the articulated language ideologies exhibited by the participant teachers in the sample.

The responses to the number patterns questions in the *second part* of the study data were captured for quantitative data analysis. Coding of the language use evidenced in the responses to the items on patterns was undertaken by the multilingual research team. As discussed in the theoretical orientation two language ideologies prevail with regard to language use: Monoglossic ideologies (belief in the value of pure language use—that only one pure language should be used to express ideas and give explanations) and heteroglossic ideologies (belief in the value of mixed language use—a pluralist (mixed) use of the full language repertoire of a speaker). Codes were drawn up to bring out the essence, in terms of language use, of the written responses to the mathematics number patterns items, allowing normally qualitative data from part two to be captured as quantitative data for descriptive statistical analysis. Codes for language use were assigned to indicate pure (IsiZulu/Setswana/English) or mixed language use (TswaEng/EngZul) so that insight into embodied language ideologies could be discussed.

Coding workshops were held to enable consensus on how to code particular responses (*part one*) and explanations (*part two*) after which data entries and codes were entered into excel spreadsheets designed for the data capture by the research team. The correct translations given in the *third part* of the study data were all entered into the excel sheets. Spelling was not penalised and appropriate symbolic representations or drawings were accepted as correct. Quantitative analysis of this data (using word counts of correct responses) was used to give insight into knowledge of the mathematical register of patterns and the range of terminology used by the participants.

6   Findings

6.1   Home Language and LoLT

The first research question addressed the issue of the relationship between the LoLT (a policy determinant) and the home language of the teachers and students in the sample schools. The LoLT of a school is chosen in consultation with the School Governing Body and more than one LoLT may be elected for use at a school. This is how the curriculum policy is thought to enable the multilingual education system envisaged by the South African LiEP (DBE, 2010). However, although there were three LoLTs in the 20 schools that participated in the study, the teachers spoke seven languages between them and there were 13 home languages spoken by the students in the sample. Even if according to the curriculum policy schools may choose more than one LoLT, as it can be seen from this data, this choice cannot always satisfy all requests by parents and students (see Table 1). This is particularly the case in the Gauteng Province and other urban contexts, due to the practical implications of multilingual communities.

A key problem in terms of genuine multilingualism in schools is that according to policy (which is monitored by higher level officials), only the chosen LoLT of the school may be used in classes. The data from this research indicate that policy is not serving the multilingual population since the majority of teachers (90%) and many students (64.3%) indicated that they spoke at least three languages. The match between LoLT and home language was higher for Grade 3 teachers, which makes sense in relation to the LoLT policy but in Grade 4 the LoLT (for all schools) is English and there were only five Grade 4 teachers in the sample who indicate English as their HL. Similar tables were drawn up for the students in the sample and the correspondence between the student home language and LoLT in the sample schools was also low (37% IsiZulu and 11% Setswana). Teachers are professionals and no doubt they accept jobs in schools where the LoLT is their HL, or is a language in which they are proficient, but the findings show the current system does not draw on the multilingual resources of the teacher and student population.

The relationship between teacher and student home language were even more stark as shown in Table 1.

As can be seen in Table 1, the match between teacher and student home language in Grade 3 classes was 45% for IsiZulu speakers and 7% for Setswana speakers while in Grade 4 classes it was 21% for IsiZulu speakers and 10% for Setswana speakers. In both Grades 3 and 4, for English, the match was 0% (Sapire, 2017). The table also shows greater variation in home languages spoken by teachers and students in Grade 4. These findings show that the current system of monolingual LoLT selection does not yield the desired policy outcome of teachers teaching and students being taught in their home language.

Evidence from the second and third parts of the study data gives insight into the language use of teachers and students, revealing clashes between the monoglossic ideology embodied in the curriculum policy document (which supports monolingual LoLT selections) and the heteroglossic language ideologies of some teachers (and students). In the next section of this chapter, we report on the findings on language use, evidenced in the highly multilingual study context.

6.2   Language Use in Written Mathematical Explanations

Firstly, we discuss findings in relation to the question, *What variation in language use is evident when teachers and students write mathematical explanations?* Close to 90% of all teachers and students responded to all questions and gave explanations

Table 1 The relationship between teacher and student home language: Grades 3 and 4. (Gr 3 n = 1506; Gr 4 n = 1352)

Learner HL/Grade	Teacher HL													
	English		IsiXhosa		IsiZulu		Sepedi		Sesotho		Setswana		Xitsonga	
	Grade 3 (%)	Grade 4 (%)	Grade 3 (%)	Grade 4 (%)	Grade 3 (%)	Grade 4 (%)	Grade 3 (%)	Grade 4 (%)	Grade 3 (%)	Grade 4 (%)	Grade 3 (%)	Grade 4 (%)	Grade 3 (%)	Grade 4 (%)
Afrikaans	0	0	0	0	0	0	0	0	0	0	0	0	0	0
English	0	0	0	0	1	1	0	0	0	0	0	0	0	0
IsiNdebele	0	0	0	0	2	1	0	2	0	0	0	0	0	0
IsiSwati	0	0	0	0	2	1	0	0	0	0	0	0	0	0
IsiXhosa	2	2	0	0	5	3	0	0	1	2	1	1	0	0
IsiZulu	4	4	2	3	45	21	0	4	2	17	3	5	1	3
Sepedi	0	1	0	0	1	1	0	0	0	0	3	4	0	0
Sesotho	1	0	0	0	5	2	0	0	0	2	2	2	0	0
Setswana	3	1	0	3	0	0	0	0	1	0	7	10	1	0
Tshivenda	0	0	1	0	0	1	0	0	0	0	0	0	0	0
Xitsonga	0	0	0	0	2	1	0	0	1	1	0	0	1	0
Arabic	0	0	0	0	0	0	0	0	0	0	0	0	0	0
Shona	0	0	0	0	0	0	0	0	0	0	0	0	0	1
Total	10	8	3	6	63	32	0	6	5	22	16	22	3	4

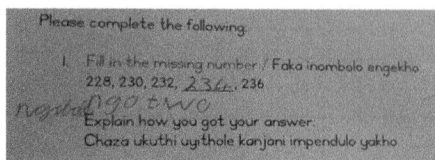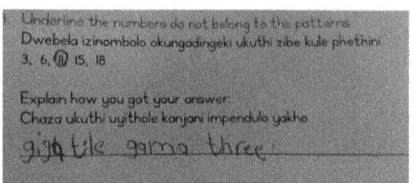

Fig. 2 Examples of mixed language used in explanations

for their answers. The language use analysis, carried out on all responses and explanations, was thus carried out on a large dataset which had been rigorously prepared and cleaned. In multilingual classes, there are many different languages that teachers and students may call on as a resource when answering mathematical questions. These may be the same as or different from the LoLT of the classroom. Knowledge of mathematical content is extremely important (the goal of mathematics teaching is mathematics learning), but the purpose of this study was to give the students (and teachers) an opportunity to speak about mathematics by giving reasons for how they worked out their answers. The explanations they wrote were used to gain insight into their language use when writing mathematics. Teachers and students used both verbal language and mathematical symbols in their explanations and they also used pure and mixed language to say what they wanted to say. Examples of explanations are shown below, selected to show a range of possibilities according to the codes of mixed and pure language use. Figure 2 gives two examples of mixed language use in student explanations.

In the first example in Fig. 2, IsiZulu is used as the main language in a mix with English, which is used to write the number word. In the second example a mix of Setswana and English occurs in a similar way. This was one of the most common forms of language mixing found in student explanations. This is a classic 'mix' which teachers wish was allowed (seen in the responses to the open-ended questions which are discussed later in the findings), evidence of heteroglossia in practice, officially disallowed by CAPS policy. There are officials who attempt to stamp out such mixing, in favour of pure language use conforming to a monoglossic language ideology (Sapire, 2018). There were other variations in language mixing: some students used both English and Setswana/IsiZulu in their explanations but in full sentences. For example, they might explain Question 1 in IsiZulu and Question 3 in English. Some students wrote their full explanations in two languages without mixing languages—for a particular question they gave an explanation in both English and Setswana. Figure 3 gives two examples of pure language use in mathematical explanations.

In the two different examples shown in Fig. 3, only one language is used in the explanation, which was the code for pure language use. The first example is from a Grade 3 response, with Setswana used for the explanation and the second one is an English explanation, given by a Grade 4 student.

Figure 4 shows the spread of language use in the explanations given by teachers

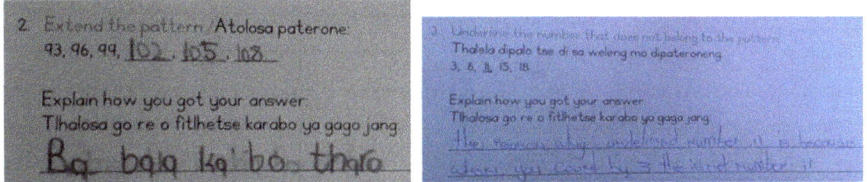

Fig. 3 Examples of pure language used in explanations

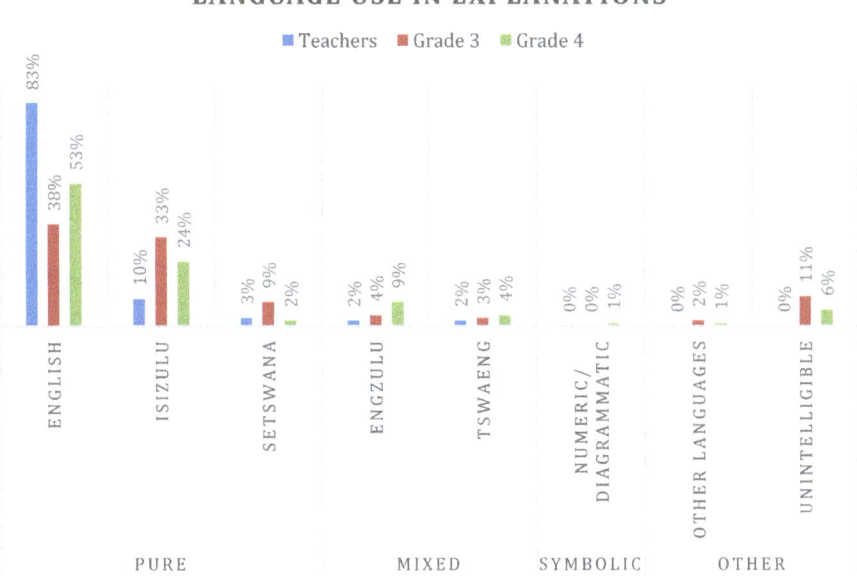

Fig. 4 Language of expression in teacher and student explanations

and students. Three categories of language use are identified in the figure: pure, mixed and symbolic.

The findings show that for both teachers and students pure language use was most common when giving explanations. Figure 4 clearly shows that most of the explanations were given in a pure language. In that category, English was most highly represented for both teachers (83%) and students (Grade 3—38% and Grade 4—53%) even though English was not the home language of the majority of the participants. This is evidence that, in spite of the multilingual context, the monolingual LoLT system enforced by CAPS policy pervades and strongly influences language use. The percentages for student explanations in all sub-categories other than English were higher than the percentages for teachers. The use of IsiZulu and English together was coded as EngZulu and the use of Setswana and English together was coded as TswaEng. There are interesting differences in these categories between Grade 3 and

Table 2 Teachers' language use preferences with reasons ($n = 62$)

Theme	Percentage
I use mostly English (it is the common student language or it is the LoLT)	85
I code-switch to support students' understanding	52
English is better for the teaching of mathematics (terminology advantage).	30
I use mostly IsiZulu/Setswana (it is the LoLT)	26
IsiZulu/Setswana is better—it is the home language and better understood by students	24
English should be used because of the policy switch to English in Grade 4	17
Teaching maths in IsiZulu/Setswana is difficult (terminology disadvantage)	9

Grade 4 students. Grade 3 students used more indigenous languages than Grade 4s. Language mixing was not common, but was higher in the Grade 4 groups than in the Grade 3 groups. Symbolic language (broader use of the mathematics register) was also not common and was used more by Grade 4s which can be explained since Grade 4s would have greater knowledge of the mathematics register.

Aligned with the finding in relation to more pure language use in explanations can be seen in Table 2, that most of the teachers favour one language when they teach and that this language is determined by the LoLT.

The CAPS policy thus appears to be influencing teachers' language use and creating a system that favours multiple monolingualism rather than multilingualism. A high percentage of teachers (85%) expressed a preference for English as the LoLT. This could be seen as policy compliance, since from Grade 4 onwards, English is the LoLT in all schools in spite of it not being the home language of the majority of students. Despite the general tendency towards the use of pure language, 52% of teachers indicated that they used code-switching to support student understanding showing that while their expressed preference is evidence of a heteroglossic ideology, the enactment of this is monoglossic. 30% of teachers mentioned the terminology 'advantage' of English while 9% spoke of the terminology 'disadvantage' of the indigenous languages. Number names are worth noting here as they are mentioned by many teachers, who say things like, *It is difficult to write number names in IsiZulu or Sesotho.* This leads into the next section of the chapter which looks more closely at the mathematics register.

6.3 Language Use with Regard to the Mathematics Register

Secondly based on the third part of the dataset, we discuss findings in relation to the question, *What variation in language use is evident when teachers and students translate selected mathematical terminology (words and phrases) on the mathematical topic of patterns?* Some of the words/phrases could be translated using non-verbal responses—examples of these are given in Fig. 5 before the statistical analysis of the

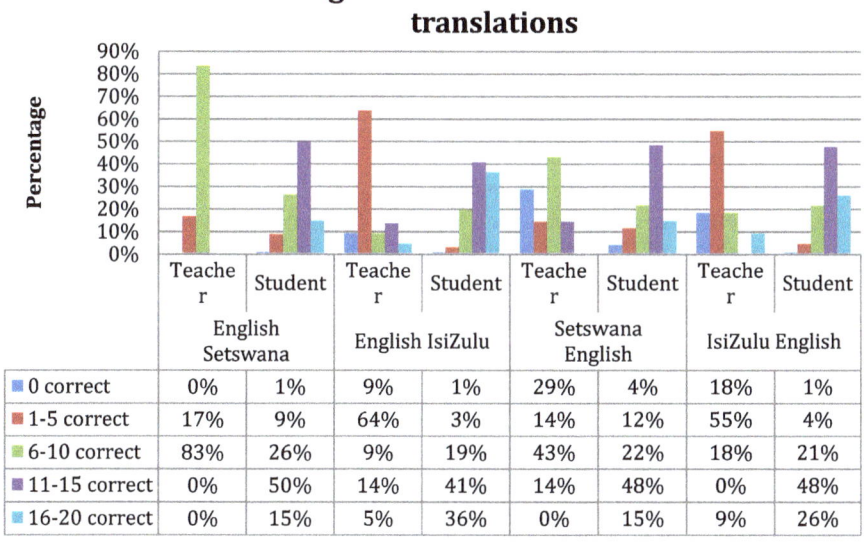

Fig. 5 'Translations' given diagrammatically, symbolically or numerically

translations is given.

Figure 5 shows three such types of words/phrases. Firstly, the operation names (used when describing rules for numerical patterns) were given symbolically by some students (e.g. subtract shown as—), secondly some words/phrases were given using numeric examples (e.g. number pattern shown as 4, 8, 12) and thirdly, words that could be visualised were given diagrammatically (e.g. geometric pattern, circle, square and triangle). In terms of the mathematics register, all these were accepted as correct 'translations' between languages. The findings for the translations (given as percentages) according to language and direction of translation for teachers and students are shown in Fig. 6.

It can be seen from Fig. 6 that while on average 15% of the teachers were able to translate between 16 and 20 words correctly almost half of teachers (49%) translated between 11 and 15 words correctly. This was better than the students, who as it can be seen in Fig. 6, struggled to translate the mathematical words. Only 1% (33 out of 2981) students were able to translate between 16 and 20 words/phrases correctly.

Percentage of correct teacher and student translations

	Teacher	Student	Teacher	Student	Teacher	Student	Teacher	Student
	English Setswana		English IsiZulu		Setswana English		IsiZulu English	
■ 0 correct	0%	1%	9%	1%	29%	4%	18%	1%
■ 1-5 correct	17%	9%	64%	3%	14%	12%	55%	4%
■ 6-10 correct	83%	26%	9%	19%	43%	22%	18%	21%
■ 11-15 correct	0%	50%	14%	41%	14%	48%	0%	48%
■ 16-20 correct	0%	15%	5%	36%	0%	15%	9%	26%

Fig. 6 Percentages of correct teacher and student translations by direction of translation

Table 3 Comparison of translation success according to LoLT and direction of translation

| | Comparison: percentage differences | | | | |
	16 < t < 20	11 < t < 15	6 < t < 10	1 < t < 5	t = 0
3Z(-E)-EZ	0	−7	−11	7	13
3Z(-E)-ZE	0	−1	−2	16	11
4Z(-E)-EZ	2	5	11	−9	−10
4Z(-E)-ZE	1	0	−5	−5	9
3S(-E)-ES	1	14	2	−37	16
3S(-E)-SE	6	1	6	−8	− 5
4S(-E)-ES	1	4	30	−17	−22
4S(-E)-SE	3	14	6	−27	7

The most common number of correct translations by students was between 1 and 5 (of 20) words correctly translated. There were many students who could not translate any words correctly—more so for IsiZulu than for Setswana.

As discussed above, matches for home language and LoLT for students were very poor. An investigation of the translation success across the eight different translation sets revealed certain patterns. This is shown in Table 3.

Table 3 gives the percentage differences between translation success in schools with an indigenous language LoLT and English LoLT according to the direction of translation. The positive differences show, admittedly for very low success rates, strength in the indigenous LoLT schools. The negative differences show the strength of translations in the English LoLT schools, for all grades and directions (apart from the Grade 3 IsiZulu group). Although these differences are noted, there is not one strong pattern. Success in the translation activity was so low for students, it could be inferred that the CAPS policy, in pushing for multiple monolingualism creates problems for speakers who may know some maths words in one language and not in another. This reduces their power of expression, since it does not allow speakers to use their full language repertoire.

It is of interest for the teaching and learning of mathematics to note which of the words were better translated and which were not. The pattern of successful translation was similar across all of the translation sets. Counts of all correct translations were recorded and represented graphically, one example of which is shown below in Fig. 7.

As it can be seen in Fig. 7, commonly used words that are used across more mathematical content areas than that of patterns were translated better by both students and teachers. The words that were more often correctly translated were: add, subtract, fives, counting in 10s, circle, triangle and square. Most students (and some teachers) were not able to translate the words flow diagram and interval. Other words that were not well translated (or translated with mixed success) by both teachers and students were: counting on, forwards, describe, multiple, number sequence and geometric pattern.

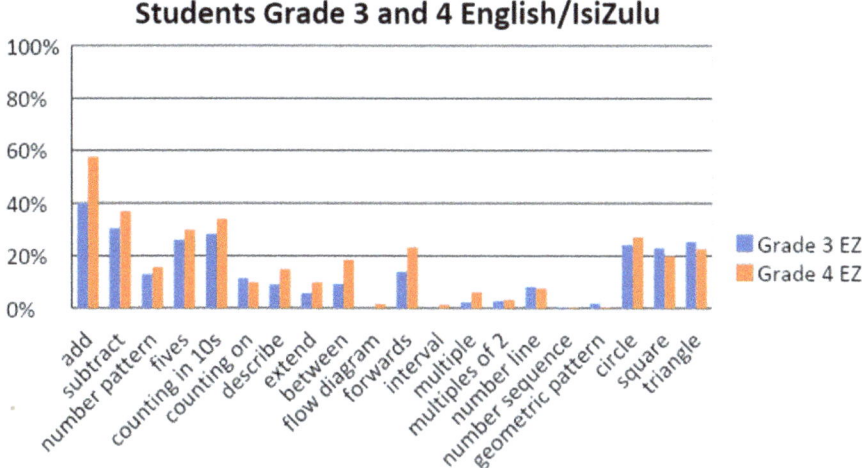

Fig. 7 Percentage correct Grade 3 and 4 student translations: English-Zulu

The variation and accuracy of the translations give insight into teachers and students knowledge of what could be called the 'standardised' vocabulary of patterns. The finding is clear—teachers' knowledge of the vocabulary is far from perfect and student knowledge is extremely poor (see Figs. 6 and 7). Overall (more so in the student sample) many variations in spelling and several synonyms were found. The synonyms (especially in the indigenous languages) should be incorporated into the Department of Arts and Culture (DAC) dictionary, since translators tend to restrict themselves to using the words in the DAC dictionary. The translation activity revealed issues could arise in regard to language use, especially in written texts if the texts use only the more 'pure' word forms recommended in the DAC. For example, the DAC recommended word for number pattern is the more formal *popegopalo* while the less formal word for pattern viz. *diphetheneng tse tsa dipalo* which is used in student print material might be more familiar to some. Our findings support that the DAC list is not adequate as a standard for translations of school learning material particularly since many dialects exist which do not all use the same words (Mojela, 2008). All possible alternatives for words should be included in general word lists so that they are more representative of the spoken languages. This and other research in South Africa indicates that language is not highly standardised (Mojela, 2008; Bokamba, 2014) although the CAPS policy has assumed that it is.

7   Discussion and Conclusion

The study was carried out to investigate the reality of the implementation of the South African Language in Education Policy. The sample schools provide evidence of a system of multiple monolingual options which do not provide fully for the multilingual population they serve. As evidenced in the findings there is a poor relationship between teacher home language and LoLT in the sample schools and the correspondence between teacher and student home language was even lower. This means that some teachers are teaching and many students are learning in a language that is not their main language. The literature has shown that in multilingual contexts, opportunities that provide for multilingual learning promote understanding (Makalela, 2015; Planas, 2018). Hence we argue that the purist ideology driving the CAPS policy which maintains that one exclusive LoLT should be used in a class may favour language use that could compromise mathematics learning and teaching since given the context in South African schools, particularly in urban areas, this would result in a LoLT that does not align with the student population.

There are two key findings that mitigate against meaningful implementation of CAPS policy. The first is that the LoLTs of the sample schools do not correspond well with the home languages of the teachers and students and the second is the evidence of a poor working knowledge of the mathematical vocabulary, more so in the two indigenous languages (IsiZulu and Setswana) that were the focus of this study. This chapter used a three-part dataset to shed light on the extent to which multilingualism is being enacted in schools in a context where it should be plain to see but findings showed teachers and students using more pure language than mixed languages. This is not surprising in the context of a policy that promotes the use of pure language. Added to this, we found that language use did not align well with the chosen LoLTs of the schools. What was seen is that English is being used in favour of other languages, most likely because there is a commonly held belief that English should be used for the teaching of mathematics (Setati, 2008). Although English is the home language of a very small percentage of the participants in the study it was reported as a second (or other) spoken language by 74% of the teachers and 68% of the students in the sample. Since language proficiency in any one of the languages is not guaranteed, as was shown in the translation activity, the use of a single language (the current endorsement of the South African CAPS) for the teaching of mathematics might not offer the optimal learning opportunity. This study shows that in multilingual contexts there is evidence of speakers drawing on more than one language of expression (even when they are not meant to be doing so) and translanguaging (García & Wei, 2014) for the learning and teaching of mathematics needs to be considered. Most teachers in this study reported that they speak at least four languages and it is common practice for teachers to teach in a language that is not their HL. In such a rich multilingual context, teachers and students should be allowed to draw on their multilingual resources and the evidence found in this study shows that they already do, despite policy constraints.

If teachers' language use could be used to infer their language ideologies, in relation to flexibility of language use, there is evidence of both monoglossic (purist) and/or heteroglossic (pluralist) ideologies. Purist language use requires standardisation of the lexicon and broad knowledge and use of this lexicon. The main arguments on the two sides of the purist/pluralist debate are: purists would say that there is a need to know all the precise words and definitions in order for conceptual discussions to be meaningful while pluralists would say that language mixing is acceptable as long as there are meaning and understanding of concepts under discussion. The study suggests that standardisation of terminology is not in place and this would indicate that the implementation of a purist monoglossic policy could be problematic.

Although this research clearly paints the picture of language use in multilingual schools, and the tragedy of the policy/practice mismatch, it did not investigate the impact of language use in these schools. Further research in this area is needed in order to inform best practices for mathematics teaching at the FP level in multilingual contexts.

References

Bamgbose, A. (1999). African language development and language planning—Language and development in Africa. *Social Dynamics, 25*(1), 13–30.

Barwell, R. (2018). From language as a resource to sources of meaning in multilingual mathematics classrooms. *Journal of Mathematical Behavior, 50,* 155–168.

Barwell, R., Barton, B., & Setati, M. (2007). Multilingual issues in mathematics education: Introduction. *Educational Studies in Mathematics, 64,* 113–119.

Bokamba, E. G. (2014). The politics of language planning in Africa: Critical choices for the 21st century. In *Discrimination through language in Africa?: Perspectives on the Namibian experience* (pp. 11–27). De Gruyter. https://doi.org/10.1515/9783110906677.11.

Cai, J., Kaiser, G., Perry, B., & Wong, N.-Y. (Eds.). (2009). *Effective mathematics teaching from teachers' perspectives*. Rotterdam: Sense Publ.

Clarkson, P. (2016). The intertwining of politics and mathematics teaching in Papua New Guinea. In A. Anjum & P. Clarkson (Eds.), *Teaching and learning mathematics in multilingual classrooms: Issues for policy, practice and teacher education* (pp. 25–39). Rotterdam: Sense Publishers.

Cummins, J. (1979). Linguistic interdependence and the educational development of bilingual children. *Review of Educational Research, 49*(2), 222–251.

Department of Basic Education. (2010). *The status of the language of learning and teaching (LoLT) in South African public schools: A quantitative review*. DBE: Pretoria.

Department of Basic Education. (2014). *Report on the annual national assessment of 2014 grades 1 to 6 & 9*. Pretoria: DBE.

Department of Education. (1997). *Language in education policy 14 July 1997*. http://education.pwv.gov.za/policies%20and%20Reports/Policies/Language.

Essien, A. (2018). The role of language in the teaching and learning of early grade Mathematics: An 11-year account of research in Kenya, Malawi and South Africa. *African Journal of Research in Mathematics, Science and Technology Education*, 48–59. https://doi.org/10.1080/18117295.2018.1434453.

García, O., & Wei, L. (2014). *Translanguaging: Language, bilingualism and education*. Basingstoke, Hampshire; New York: Palgrave Macmillan.

Heugh, K. (2014). Dreams and realities: Developing countries and the English language. *Current Issues in Language Planning, 15*(1), 110–116.

Kajoro, P. (2016). Transition of the medium of instruction from English to Kiswahili in Tanzanian primary schools. In A. Anjum & P. Clarkson (Eds.), *Teaching and learning mathematics in multilingual classrooms: Issues for policy, practice and teacher education* (pp. 25–39). Rotterdam: Sense Publishers.

Kroskrity, P. V. (2004). Language ieologies. In A. Duranti (Ed.), *A companion to linguistic anthropology* (pp. 496–515). Malden: Blackwell.

Lafon, M., & Webb, V. (2010). The standardisation of African languages. Michel Lafon; Vic Webb. IFAS, pp. 141, 2008, Nouveaux Cahiers de l'Ifas, AureliaWa Kabwe Segatti. <halshs-00449090>.

Lyon, J. (1996). *Becoming bilingual: Language acquisition in a bilingual community.* Clevedon: Multilingual Matters.

Macnamara, J. (1977). *Language learning and thought: Perspectives in neurolinguistics and psycholinguistics economic theory and mathematical economics.* New York: New York Academic Press.

Makalela, L. (2015). Translanguaging as a vehicle for epistemic access: Cases for reading comprehension and multilingual interactions. *Per Linguam, 31*(1), 15. https://doi.org/10.5785/31-1-628.

Mojela, V. M. (2008). Standardization or stigmatization? Challenges confronting lexicography and terminography in Sesotho sa Leboa*. *Lexikos* 18(AFRILEX-reeks/series), 119–130.

Phakeng, M. (2017). *Mathematics education and language diversity: From language-as-problem to language-as-resource.* Keynote address at the NSTF (National Science and Technology Forum) discussion on multilingualism in STEM subjects, Emperor's Palace, Benoni.

Pimm, D. (1981). Mathematics? I speak it fluently. In A. Floyd (Ed.), *Developing mathematical thinking.* Workington: Addison Wesley.

Planas, N. (2018). Language as resource: A key notion for understanding the complexity of mathematics learning. *Educational Studies in Mathematics, 98,* 215–229. https://doi.org/10.1007/s10649-018-9810-y.

Prediger, S., Kuzu, T., Schüler-Meyer, A., & Wagner, J. (2019). One mind, two languages—Separate conceptualisations? A case study of students' bilingual modes for dealing with language-related conceptualisations of fractions. *Research in Mathematics Education, 21*(2), 188–207. https://doi.org/10.1080/14794802.2019.1602561.

Robertson, S., & Graven, M. (2018). Exploratory mathematics talk in a second language: A sociolinguistic perspective. *Educational Studies in Mathematics, Online.* https://doi.org/10.1007/s10649-018-9840-5.

Ryan, U., & Parra, A. (2019). Epistemological aspects of multilingualism in mathematics education: An inferentialist approach. *Research in Mathematics Education, 21*(2), 152–167.

Sapire, I. (2017). *Researching multilingualism in early grade mathematics: Survey—Technical report.* Johannesburg: University of the Witwatersrand.

Sapire, I. (2018). *Researching multilingualism in early grade mathematics: Interviews—Technical report.* Johannesburg: University of the Witwatersrand.

Sapire, I., & Roberts, G. (2017). *The status of the language of learning and teaching (LoLT) in schools: A quantitative overview: 2008-2016* (Draft report).

Setati, M. (2008). Access to mathematics versus access to the language of power: The struggle in multilingual mathematics classrooms. *South African Journal of Education, 28*(1), 103–116.

Trudell, B., & Piper, B. (2013). Whatever the law says: Language policy implementation and early-grade literacy achievement in Kenya. *Current Issues in Language Planning.* https://doi.org/10.1080/14664208.2013.856985.

Ingrid Sapire has been involved in mathematics teacher education for 27 years and is based at the University of the Witwatersrand. Since 2012 her focus has been on the development of materials at scale for Foundation Phase mathematics teachers and students both at a provincial and national level. All of her work in the Foundation Phase has had a focus on the multilingual project—how to best provide for students in the system, challenging the policy whenever necessary. In 2017–2018 she was the chairperson of a ministerial task team that developed a framework for the teaching of mathematics in South Africa schools. She is currently heading the multilingual Bala Wande (Calculating with Confidence) Foundation Phase mathematics programme and working on her Ph.D., in multilingualism in Mathematics in the Foundation Phase.

Anthony A. Essien is an Associate Professor and the Head of the Mathematics Education Division at the University of the Witwatersrand, South Africa. He is a series editor of the book series *Studies on Mathematics Education and Society.* His field of research is in mathematics teacher education in contexts of language diversity. He is also a current member of the International Committee (Board of Trustees) for the International Group for the Psychology of Mathematics Education (IGPME). Anthony also served as an associate editor of *Pythagoras*, the academic journal of the Association for Mathematics Education of South Africa, for 11 years. In addition to his background in mathematics education, Anthony also has a background in Philosophy.

Practices in STEM Teaching and the Effectiveness of the Language of Instruction: Exploring Policy Implications on Pedagogical Strategies in Tanzania Secondary Schools

Opanga David, Alphonse Uworwabayeho, Théophile Nsengimana, Evariste Minani, and Nsengimana Venuste

Abstract The language of instruction, particularly in STEM, has been an issue in many developing countries. In reference to Tanzanian education policy, Kiswahili is recommended as the language of instruction in primary schools (P1–P7), while English is used in all levels after the primary. This chapter argues that students need to be supported following the language pedagogical strategies to enhance the learning of subject content and the language of instruction in secondary schools. This study aimed at exploring the policy implications and contributions of language pedagogical strategies on students learning of STEM, focusing on biology. A total of 250 students were randomly selected and 36 teachers purposively selected to participate in the study. Pre- and post-tests, focus group discussions, interviews and lesson observations were used for data collection. Results showed that post-test results (mean $=$ 38.5; standard deviation $=$ 13.6) were higher than the pre-test (mean $=$ 30.2; standard deviation $=$ 12.2), in addition to significant differences in mean scores between the post-test and pre-test means (t $_{(249)}$ $=$ 5.6, $p < 0.05$). Lesson observations and teachers' interviews revealed that the strategic use of bilingual instruction increased students' interactions and activeness in class, which in turn, improved the learning

O. David (✉)
University of Rwanda-College of Education, African Centre of Excellence
for Innovative Teaching and Learning Mathematics and Science, Kayonza, Rwanda

A. Uworwabayeho
Department of Early Childhood and Primary Education, University of
Rwanda-College of Education, Kayonza, Rwanda
e-mail: auworwabayeho@ur.ac.rw

T. Nsengimana · E. Minani · N. Venuste
Department of Mathematics, Science and Physical Education, University
of Rwanda-College of Education, Kayonza, Rwanda
e-mail: tnsengimana@ur.ac.rw

E. Minani
e-mail: eminani@ur.ac.rw

N. Venuste
e-mail: v.nsengimana3@ur.ac.rw

© The Author(s), under exclusive license to Springer Nature Switzerland AG 2021   97
A. A. Essien and A. Msimanga (eds.), *Multilingual Education
Yearbook 2021*, Multilingual Education Yearbook,
https://doi.org/10.1007/978-3-030-72009-4_6

of content through the English language. Further, students viewed mixing Kiswahili with English as more helpful and wish to see this approach adopted in teaching and learning. This book chapter, therefore, suggests that the education policy needs to acknowledge language pedagogical strategies combining Kiswahili and English to allow students to effectively learn STEM, particularly biology.

Keywords Language of Instruction · Language policy · Pedagogical language strategies · STEM

1   Introduction

Language in education policy has been a matter of concern in the swiftly increasing systems of developing economies (Brock-Utne, 2014; Samuelson & Freedman, 2010). For most countries' education systems in Sub-Saharan Africa, English language has never been the first language for both teachers and students (Brock-Utne, 2014). The evidence is that English as a language of instruction is introduced in middle or upper primary education. The reason behind this late use of English as a language of instruction is that learning in a familiar language is encouraged in early education to raise academic achievements at an early age rather than in the foreign language. However, the status quo has been challenged in support of multilingualism in education (Rubagumya, 2003).

Viewing the matter in the East African region; Tanzania, Uganda and Kenya, in particular; these countries were the former British colonies, where English has been a language of instruction in the education system since the colonial period (Abdulaziz, 1982; Rubagumya, 2003; Kyeyune, 2003). In the region, Rwanda was under the Belgian colony, hence used French as a language of instruction until 2008 (Gahigi, 2008), then adopted English in teaching and learning since 2009 as the country joined the East African Community in 2007 and commonwealth by 2009. In Rwanda, students' home languages are used as the language of instruction with English only taught as a subject in nursery schools (Rwanda Education Board [REB], 2015a, 2015b, 2020), and then all subjects are taught in English from primary one to the higher learning levels. The common characteristic in East African countries is that a small section of the population speaks English outside of school and office activities (Samuelson & Freedman, 2010; Rubagumya, 2003; Abdulaziz, 1982; Kyeyune, 2003).

In Tanzania, following World War I, the British government took over the administration of German East Africa. Kiswahili was therefore conserved as the language of instruction from primary 1 to primary 5, and English was used for the last three years of primary and post-primary (Rubagumya, 1990). During the colonial time, English was the official language of colonial administration, and it was used to train a minority of elite Tanzanians to assist in British colonial administration (Roy-Campbell, 2001). Later in 1965, the education structure changed to 7 years of primary education. In

1967, Kiswahili became the official language of instruction in primary, then English was used in post-primary education (Rubagumya, 2003; Vavrus, 2002).

Further, East African countries consider Science, Technology, Engineering and Mathematics (STEM) subjects as important to enable citizens to manage the environment and to contribute to the development of the nation (Tanzania Ministry of Education, Science and Technology [MoEST], 2015; REB, 2015a). In Tanzania, STEM subjects are taught from year three of primary education. At the secondary school level, all students learn physics, chemistry, biology and mathematics until the second grade known as the second year of ordinary secondary school level (MoEST, 2015). After, all students continue to study biology and mathematics as compulsory subjects to the fourth year of secondary education. Thereafter, students are free to choose between physics and chemistry in the third year and fourth year of secondary education. Successful candidates are expected to leverage different fields of science and technology in colleges of higher education and vocational training (MoEST, 2015).

Despite the case, teachers and students are challenged by integrating STEM subject content and English language in teaching and learning in Tanzania (Gabrieli et al., 2018; Barrett et al., 2014). Besides English, Kiswahili is the lingua franca of the country, used as the language of instruction for public primary schools since 1965. The Tanzanian education policy specifies that the language of instruction in primary must be Kiswahili, while English has to be used in post-primary education. On the other hand, the English language is used in few pre-primary schools referred to as English medium pre-primary and primary schools (Ministry of Education and Vocational Training [MoEVT], 2014).

Nevertheless, in light of supporting students from diverse backgrounds of more than 120 ethnic communities who are not conversant in English (Galabawa, 2006), the Tanzanian government took initiatives to introduce a new policy to address various language educational challenges. The Tanzania education and training policy (ETP) approved in 2014 (MoEVT, 2014) states that Kiswahili has to be used for teaching and learning at all levels of education and training. The government would then put in place a mechanism to make the use of this language sustainably and effectively to provide effective national and international education and training. Besides, the Government was committed to continuing with the process of enhancing the use of English in teaching and learning at all levels of education and training. Unfortunately, the 2014 ETP is yet to be put into implementation as the 1995 ETP reformed in 2005 still holds to date (Gabrieli et al., 2018). Even though the policy clarifies the use of Kiswahili and English, there is still a challenge on how effectively the language of instruction recommended by the policy can be well articulated to ensure effective teaching and learning for STEM subjects.

Despite attempts done to improve the status quo in the country, students still face difficulties using English as a language of instruction in secondary schools. Consequently, they acquire very little knowledge in STEM (Barrett & Bainton, 2016; Gabrieli et al., 2018). A quick transition from primary to secondary education without clear language support can be among the reasons for the limited acquisition of knowledge in STEM as has been indicated by Gabrieli et al. (2018). Therefore, there is a need to explore the policy implications and the contribution of language pedagogical strategies on students learning STEM in Tanzania. This study seeks to fill the gap and answer the following research questions: What are the language policy implications on pedagogical strategies in teaching and learning STEM? How can effective pedagogical language strategies supported by bilingual instructions be used in teaching and learning STEM? The findings of this study would illustrate the link between policy interpretation and the choice of pedagogical language strategies in Tanzanian secondary schools.

To answer these research questions, we conducted STEM research using biology subject, the content of invertebrate systematic. Systematic biology has been preferred because it possesses subject-specific terms in foreign languages, mainly Latin, that is internationally accepted as the language of systematic biology. Further, studies done by Gabrieli et al. (2018) and Ricketts (2014) indicate that among other STEM subjects, biology places critical language demands on both students and teachers. Further, biology was selected as one of the least performed STEM subjects in national examinations of the ordinary level during the past six years in Tanzania, particularly in Dodoma region, where this research was conducted (MoEVT, 2014; MoEST, 2019).

This study aimed to indicate how the combination of Kiswahili and English can contribute to solving teacher and students' language difficulties in teaching and learning STEM subjects, specifically biology, systematic of invertebrates. Results indicated positive impacts during the teaching practice, as it equipped teachers on the language strategies to support students. Further, the study informs the policy-makers and curriculum developers that a unified form of strategies to effectively teach and learn STEM subjects is missing from the current MoEST education policy, particularly the use of bilingual classroom instruction. Furthermore, the results of the present research would be useful to education stakeholders, school heads and STEM teachers, who may wish to advance the integration of content and the use of the English language in teaching and learning STEM.

2   Literature Review

2.1   Education Policy Reforms in Tanzania

Since independence in 1961, the government of Tanzania attempted numerous restructures of the educational system to meet the development goals and objectives to generate desired outcomes (Cooksey, 1986). It can be argued that the history of Tanzania's educational system is a composite one, and has been driven by many goals, ideologies, intentions and motives (Nyerere, 1968). The early attempts of reforming education and training policy was that of 1967 marked as education for self-reliance policy (ESR). The ESR was a work-oriented and rural-oriented vocational education (Nyerere, 1968).

Later in 1995, the education and training policy (ETP) was introduced (Gabrieli et al., 2018). The shift from ESR to ETP was marked by significant movement from vocational education to broader education policy. The ETP was accompanied by a transformation of policy stress from rural-oriented vocational education to technology education. The ESR and ETP are considered to be major policy reforms ever made by the Tanzanian government. However, there are minor reforms such as that of 2005, marked by the shift from content to competence-based curriculum. Later in 2014, another reform was made and focused mainly on attaining the national visions including that of 2025. This was highly driven by the national philosophy of industrialization, putting more emphasis on science and mathematics, technological education and language of instruction (MoEVT, 2014). This shift was necessary for building the twenty-first century skills for the shift to middle income and knowledge-based economy.

The current education and training system in Tanzania is 2-7-4-2-3+. This means 2 years of pre-primary education, 7 years of primary, 4 years of secondary ordinary level, 2-years of secondary advanced level, and a minimum of 3 years of the university and higher learning institutions (MoEVT, 2014). Primary and ordinary levels are considered basic education, and compulsory for all students. However, after a thorough review of the education and training policy of 1995, the 2014 education and training policy suggested a more feasible structure of 1-6-4-2-3+ (MoEST, 2015) to enable students to complete the formal education cycles in a short period. Unfortunately, this is not yet put into implementation as articulated.

2.2 Language Supportive Pedagogy and STEM Teaching and Learning

In most Sub-Saharan countries, English is used as the language of instruction in education systems (Brock-Utne, 2014). Studies (Brock-Utne, 2014; Samuelson & Freedman, 2010) showed that teachers and students have difficulties using English while teaching and learning, as this is not a native language to them. One of the approaches introduced to overcome challenges imposed by the use of English in teaching and learning is the content and language integrated learning (CLIL) approach. This is a dual-focused educational approach in which an additional language is used for the learning of the content and language (Coyle et al., 2010). The CLIL was launched in Europe around the 1990s. Later, it was either adapted or adopted by some European, American, Asian and African countries (Barrett & Bainton, 2016; Dalton-Puffer & Smit, 2013).

In the context of CLIL, an additional language could be any language such as foreign, second, or minority languages (Marsh, 2002). In other words, it may refer to any language other than the first language. In some cases, CLIL is limited to a foreign language and can take place at various levels of education such as pre-school, primary school, secondary school, and post-secondary education. In STEM, the CLIL aims at supporting students to learn subjects taught and assessed using English as a medium of instruction (Massler et al., 2014). Massler et al. (2014) argue that the use of CLIL depends on students' and teachers' profiles, targeted languages, a balance between content and language of instruction, and other pedagogical issues such as teaching and learning material development and instructional methods. In Tanzania, several pedagogical approaches have been introduced and implemented to support students to learn STEM as a solution to policy implication (Gabrieli et al., 2018; Barrett & Bainton, 2016). These include the CLIL introduced in the 1990s, aiming at multilingualism. However, it contrasted with other transitional programs published in Massler et al. (2014) that aimed at replacing Kiswahili (L1) with English (L2), resulting in monolingualism (Mohan & Slater, 2005). Another approach was introduced by the language supportive teaching and textbook in the Tanzania project (Barrett et al., 2014). This is the language pedagogical strategy stemming from the teacher. It comprises the use of language supportive activities, that implies translating when necessary, and interpreting from English to Kiswahili for students. In this regard, teachers´ use of subject-specific language genres and glossaries would have a positive impact on students´ learning.

The CLIL and language supportive teaching and textbook (LSTT) approaches emphasize using Kiswahili to improve English (Barrett & Bainton, 2016; Massler et al., 2014). However, LSTT extends to teachers' collaboration through lesson studies, activity-based lesson development and the development of language-supportive STEM teaching materials (Barrett & Bainton, 2016). The two approaches being used to respond to policy-practice tensions created by the language in education policy, there is still a gap on how teaching and learning biology can be effectively done by using bilingual instructions.

3   Research Theoretical Context

The social constructivism learning theory proposed by Levy Vygotsky (1978) served as a reference in planning and conducting lesson observation. The theory underpins culture and language through individual interactions. It was argued that language needs to be understood by both teachers and students for meaningful learning to occur (Barrett & Bainton, 2016). Under this theory, learning practices need to be associated with students' context. In particular, social constructivists focus on social interactions and conversation patterns (Davis et al., 2000). From this perspective, students build their understanding within small groups such as pairs for example. Further, teacher–student interactions are reinforced to help students understand the content, thereafter, students are given time to present the findings.

We assumed that the theory could enable us to understand the role of policies in catalysing teachers' initiatives to create a positive environment for learning STEM subjects. Also, the classroom itself is a social environment where people having almost the same age, but with different backgrounds meet together to learn. In this regard, teachers have to organize the classroom by taking into consideration social interactions. Further, teachers have to provide support and bridge between different students' cultures, languages and social values (Wells, 1999). The theory emphasizes also collaborative learning whereby students discuss and share observations or experiences from a given set of activities.

4 Methodology

4.1 Study Area

The study was conducted in Tanzania located on the East Coast of Africa. The country is inhabited by over 55 million people (United Republic of Tanzania [URT], 2019). Administratively, Tanzania is divided into 31 regions. This study was conducted in the Dodoma region, purposively selected. It has more than 200 secondary schools and it is marked by students' poor performance in STEM. It was ranked in the least 11 performing regions for over 6 years (2014–2019). Out of 31 regions, it was recently ranked at 24th (MoEST, 2016), 19th (MoEST, 2017), and 21st place (MoEST, 2018, 2019) in form four national examinations. Data were collected from 9 secondary schools based on performance in form four national examinations. Precisely, schools with high, average, and low-performing ranks were selected for this study (MoEST, 2019). Further, school locations, specifically from urban and rural Dodoma regions were taken into consideration for school selection.

4.2 Research Approach and Research Design

This study employed a mixed-method research approach. The qualitative approach was used to explore teachers' and students' feelings about classroom practices that favour language support in teaching and learning. This was done through interviews with STEM teachers and focus group discussions with students. Further, the ex-post facto design, a quantitative approach (Kothari, 2008) involving pre- and post-classroom assessments was employed. Formative assessment was applied to evaluate the effectiveness of introduced pedagogical language strategies in teaching and learning of STEM, focussing on biology. Specifically, this study involved students from year four of the ordinary level of secondary education. The grade was selected because it is the one concerned with the topic of invertebrate systematic, Kingdom Animalia (URT, 2020).

4.3 Data Collection

Data were collected between March 2019 and August 2020. The baseline study was conducted between March 2019 and February 2020, while the intervention was done from March to August 2020. A total of 54 sessions, each of 80 minutes were observed in 9 schools. A total of 36 teachers (STEM: 27—three from each school, Language: 9—one from each school) and 250 students (28 from eight schools and 26 from the ninth school) from level four of the ordinary level participated in the study. The sample size of students was chosen based on the recommended class size in Tanzania which should not exceed 40 students (MoEST, 2019), while that of teachers depended on the number of biology and language teachers available at the school level.

The study was divided into three stages: lesson preparation, lesson implementation and lesson observation. Before each stage, a consent form was signed by teachers and students. When a student was less than 16 years old, ethical clearance was issued from parents. Further, participants were assured about confidentiality and guaranteed that data have to be only used for research purposes.

Stage 1: Pre-test and Lesson Preparation

This stage consisted of the lesson observation and formative assessment in biology focussing on systematic invertebrates. Data were collected using the interviews, observation checklist, marking the evaluation sheets and marks recording. Marks and lesson observations were used to formulate the intervention to help teachers and students to overcome identified challenges. The intervention consisted of training participant teachers and sharing the experience about the inquiry and strategies that can be used to support students with English language difficulties. Further, 20 teachers purposively selected, were interviewed before the training to get their views about supporting students with English language in the context of learning invertebrate systematic. After the training, the 80 minutes biology lesson was prepared by improving the one observed, where the subject content and learning resources were revised and improved.

Further, the elements of language objectives were introduced as a new entity indicating lesson activities that would support students to improve the language of instruction and understand the lesson content. The language strategies focussed on the use of language supportive activities, translate when necessary, interpret for students, help students to pronounce correctly, use of language genres specific to the subject and topics, provision of a glossary, use of simple English sentences and bilingual instructions where Kiswahili and English could be used strategically to concurrently learn. Further, the research team recorded data on changes in the revised biology lesson plans and information obtained from experience sharing using notebooks.

Stage 2: Lesson Implementation and Lesson Observation

This consisted of teaching the lesson prepared in stage one. One of the teachers who participated in the lesson planning volunteered to teach, while others remained the observers. During the teaching and learning processes, students were given equal opportunities to learn and interact using both Kiswahili and English. During the group discussion, students were allowed to use Kiswahili and report in English.

Observers used a predesigned checklist to assess the effectiveness of pedagogical language strategies indicated in the lesson plan. Teachers' interactions with students throughout the lesson were noted, the ability of students in speaking and reading skills were recorded, and challenges faced by teachers were also noted.

Stage 3: Post-lesson Evaluation

This stage consisted of the evaluation of the success of the lesson. Teachers and the research team reflected on successful points implemented as planned in the lesson; discussed the points that need improvement; and reflected on how the pedagogical language strategies supported the teaching process. To explore differences experienced by students while teaching and learning using English only, and using both English and Kiswahili, the interview was organized with 90 randomly selected students and the focus group discussion was organized with 10 students at the school level. Data on the advantages of teaching using both Kiswahili and English were also collected through the interview with 20 teachers who taught the lesson.

At the end of the lesson, all students who attended the revised lesson were assessed using the content-based questions, prepared following Bloom's taxonomy to verify the achievement of the objectives. Scripts were collected and marked focussing on the understanding of biology content and the English writing skills. The passing mark was fixed at 50%. To avoid biases in marking, scripts were checked by each member of the research team and then verified by a private person selected out of the research team, but having skills in assessment. Results were then recorded and compared with those of the pre-test to evaluate changes in performance. Finally, each student's English writing, reading, speaking and listening skills were checked over the set of given writing and reading class session activity.

4.4   Control of Threats to Internal Validity

Based on Maxwell's (2012) recommendations, we planned for intensive and long-term research study working with the same observers and participants over the systematic of invertebrate, subdivided into six class sessions and six subtopics. To ensure that data do not apply to only one observation or one sample population, multiple sources of data namely, students' evaluation sheets, focus group discussions, and interviews with students and teachers were used. Further, Bloom's taxonomy guided the development of the assessment questions for each session. Nevertheless, observed teachers were selected under conditions of being either a biology or English teacher, and having at least 6 years of teaching experience in secondary schools.

4.5   Data Analysis

Means and standard deviations were used to compare changes in student's performance between pre- and post-tests. The paired sample *t*-test (Kothari, 2008) was used to check significant differences in performance before and after intervention

Table 1 Assessment of writing, reading and speaking skills adapted from Shohamy et al. (1992) and Knoch (2011)

Structure of paragraph	Ranking category	Ranking scale
a. Writing skills assessment		
No errors in terminology, sentence structure and morphology	Correct	4
No errors in terminology and sentence structure, but with fairly correct morphology	Fairly correct	3
No errors in terminology and morphology but with errors in sentence structure	Poor	2
No errors in morphology but has errors in sentence structure and terminology	Very poor	1
Errors in morphology, sentence structure and terminology	Not correct	0
b. Reading and speaking assessment		
Reading/speaking had the exact pronunciation, break in proceedings and fluency	Correct	4
Exact pronunciation without respect pauses and fluency	Fairly correct	3
Reading adhered to pauses but did not follow the exact pronunciation and fluency	Poor	2
Reading was totally not following the exact pronunciation, break in proceedings, and fluency	Very poor	1
Total failure to read the sentence	Not correct	0

for the effectiveness of pedagogical language strategies and the understanding of the lesson content. Further, frequencies and percentages were used to test for the extent to which teachers used pedagogical language strategies and student's English language skills evaluation. Thematic analysis was employed in the analysis of qualitative information, and themes were developed based on the data generated. Data were analysed using SPSS Version 16.0 software. Further, the writing and reading skills consisting of categorical variables related to students' performance were set to standards as per the five-point rating scale (Table 1).

5 Results

5.1 Implications of Education Language Policy on Teaching and Learning STEM

Results collected from stage one indicated that only 10.0% of teachers were able to set language objectives to be achieved in class. Unfortunately, these objectives were not clear enough to support students to understand biology subject content. Further, concerning the activities set for students to be done during teaching and learning, only 25.0% reinforced the subject competences, and the majority of them (75.0%) were teacher-centred. While the policy dictates the use of English at this level of education, a substantial number of teachers used Kiswahili throughout the lesson with minimal activities to reinforce students' subject competences.

Before the intervention, students had the following views: the captured evidence below is part of respondents' responses:

...English language is much used, but the majority of us are not competent in English. It is very hard to understand when teachers are teaching biology using English only. To solve the problem, teachers tend to use both, English and Kiswahili, to help us to understand the lesson... sometimes teachers use Kiswahili throughout the lesson. (Student from school 2)

This was also testified by teachers when asked about how language in education policy creates tensions in the teaching and learning of biology:

...Yes, the policy requires teachers to teach in English, but when we are teaching in English, students fail to understand the subject content. We are receiving students from diverse language backgrounds, so that we have to use Kiswahili and other local languages to support them. (Teacher from school 4)

After the intervention, the interview with teachers and focus group discussions with students revealed that mixing both Kiswahili and English was helpful. This is shown by the parts of captured evidence from students and teachers:

... The knowledge and skills from the training to support students learning biology were helpful. It is now easy for me to give activities according to the level of students, and use different language strategies that mix Kiswahili and English, and hence help students to understand the biology subject content. (Interview with a teacher from school 6)

... Communicating in English was very difficult for me. I am now interested to learn biology, as the teacher is now explaining scientific concepts in English and Kiswahili. (Interview with student from school 1)

5.2 Effectiveness of Pedagogical Language Strategies in STEM Teaching and Learning

Marks from pre- and post-tests conducted on invertebrate systematic biology indicated the improvement in students' performance. Statistical analysis over six successive class sessions conducted in each school indicated changes from pre (30.2 ± 12.2) to post (38.5 ± 13.6) evaluations. Further, significant differences ($t_{(249)} = 5.6, p < 0.05$) were found between the marks obtained in pre- and post-evaluation. Besides, it was found that when students discuss in Kiswahili and report in English, there is a change in students' interaction, improved academic reading and writing skills, and mastery of the subject content (Table 2).

The intervention helped teachers (65.0%) to fully implement teaching and learning activities as they had been planned in the lesson plan. Further, the use of mixed Kiswahili and English languages during group discussions favoured student-material interactions. A substantial number of students (49.6%) were able to clarify, interpret

Table 2 Evaluation of bilingual classroom instruction implications on pedagogical strategies

Strategy	The language used		Effect on students' learning
	Kiswahili	English	
Teacher allows students to discuss in Kiswahili and report answers in English	More often		Increased students' interactions during group work
Teacher translates where necessary	More often		Increased students' responses to the questions asked by teachers and other students
Teacher provides writing opportunities		Clear writings	Improved academic writing skills on the chalkboard and student's notebooks
Teacher provides reading opportunities		Less frequent	Improved reading skills on the chalkboard and subject notebooks
Teacher uses language supportive activities		More often	Mastery of subject through exercises provided by the teacher during the lesson
Teacher pronounces correctly as s/he is teaching the subjects		More often	Students correct mistakes by themselves
Teacher uses language genres specific to the subject or topics they teach		More often	Development of writing technical terms in relation with the subject content

and draw precise conclusions from the given activities through the use of both English and Kiswahili (Table 3). Others (55.6%) could assess themselves through individual class exercise and tests. Further, they were able to correctly answer evaluation questions and get the 50.0% pass mark and above.

Further, the use of both Kiswahili and English helped students to improve writing, speaking, reading and listening skills. Marks from the individual student's English writing, reading, speaking and listening skills assessment over the set of given writing and reading class session activity indicated that all students scored 50.0% and above compared to before the intervention, where 43.6% of students could not attain 50.0% (Table 4).

6   Discussion

6.1   Implications of Education Language Policy on Teaching and Learning STEM

Findings from interviews with teachers and focus group discussions with students revealed that the English language imposes difficulties to teachers and students in teaching and learning STEM, particularly biology. In the same position, Gabrieli

Table 3 Classroom observations on pedagogical language strategies through inquiry process (*Note* T = Teacher, S = Student, N for students = 250, N for teachers = 20)

Inquiry process phase	Frequency (S)	% (S)	Frequency (T)	% (T)	Participant (s)
Engage/excite Use of activities that stimulates students by the use of Kiswahili and English	74	29.6	13	65.0	Teacher
Explore Students discuss in groups using Kiswahili and English to foster student-material interaction	62	24.8			Student
Explain Students provide interpretations and conclude from given activities in English	62	24.8		Student	
Elaborate Teacher gives other scenarios to help students apply their knowledge in a different situation from the one they explored by fostering student–teacher interactions	52	20.8	7	35.0	Teacher
Total	250	100	20	100	

Table 4 Students' performance in English language skills (*N* = 250)

Skills	Before intervention		After intervention	
	Frequency	%	Frequency	%
Writing	50	20	60	24.0
Speaking	35	14	67	26.8
Reading	30	12	65	26.0
Listening	26	10.4	58	23.2
Performance	109	43.6	All students got 50% and above	All students got 50% and above
Total	250	100	250	100

et al. (2018) found that students struggle with the use of the English language as for the majority it appears to be the third language, the first being the mother tongue and the second being Kiswahili. Besides, Barrett et al. (2014) found that the majority of form one student could not read simple English texts well. Barrett et al. (2014) and Gabrieli et al. (2018) agree that English language skills remain a problem for most Tanzanian students. Further, the authors indicated that students in secondary schools of Tanzania are not prepared for the use of English as the language of instruction. This imposes teachers struggling with teaching the subject content in the language less understood by students. Therefore, once both English and Kiswahili become the languages of instruction, benefits shall be for both students and teachers to understand the subject content and provide clarifications of the subject concepts.

6.2 Effectiveness of Pedagogical Language Strategies in Learning STEM

6.2.1 Bilingual Classroom Instructions on STEM Teaching and Learning in Tanzania

In Tanzania, little emphasis was given to the English language as a matter that contributes to the effective teaching and learning STEM (Rubagumya, 2003; Gabrieli et al., 2018). This is added to the lack of language support to STEM teachers, specifically biology teachers. Results of this study indicated that more often teachers allow students to discuss in Kiswahili and report in English (Table 2). This is a sign that the language support using mixed language strategies has a positive impact and helps students to understand STEM content. Phillipson (1992) points out that cognitive development in Kiswahili has significant effects on English and subject content. It appears that failure to offer encouraging classroom pedagogical practices that favour literacy in Kiswahili may undermine the development of skills such as critical thinking for students. In addition, Vygotsky (1962) argues that the child can transfer the knowledge to the new language and develop a deep understanding on her/his own. In this context, Kiswahili can be used strategically to support students in understanding the subject content in English.

It was also evident that increased students' interactions, improved academic reading and writing skills, and the mastery of the subject content (Table 3). Cummins (2000) argues that in bilingual classrooms, the development of students' literacy skills can improve the understanding of the lesson. It was again observed that classroom sessions that did not value the use of bilingualism, students remained silent during the lesson. This effect may persist in students' life-time learning even when they join college and university. Puja (2003) observed that the greater part of students was silent in class. The author was impressed by the fact that as soon as the class is over, students and teachers switch over to Kiswahili and communicate freely. The authors then concluded that bilingual classroom instructions may potentially support students' understanding of the subject content by fostering classroom interactions.

On the other hand, policy and pedagogy used in Tanzania secondary schools place critical language demands on students due to insufficient strategies or teacher expertise in STEM teaching (Barrett & Bainton, 2016). Before the intervention, frequently observed features included the dominance of teacher-centred approach with inadequate student replies (Gabrieli et al., 2018). During the intervention, specific patterns of classroom interactions such as student–student interaction, student–teacher inter-action, and student–material interaction were frequently observed. We assume that the reason behind this is the training which equipped teachers with skills and knowl-edge that allowed them to link and contextualize the subject content with the use of both Kiswahili and English. This supports the conclusion of Puja (2003) about the use of bilingualism while teaching and learning.

Furthermore, Rubagumya (2003) supports that English as a medium of instruction constitutes a problem of education for the majority of the Tanzanian people, to which English is less used by the population of Tanzania. This affects the use of English as a language of instruction since it hinders effective communication among students and between teachers and students. This indirectly slows down the students' speaking skills, thus remaining silent and less confident in class. In essence, the language of instruction policy needs to be well articulated to support its effective implementation in teaching and learning STEM in Tanzania.

6.2.2 Students' English Writing and Reading Skills Development in STEM Teaching and Learning

Writing and reading skills are very important for students to develop an understanding of curriculum concepts (Cummins, 2000). In Tanzania, English being the language of instruction in secondary schools, writing and reading are important for students to understand STEM concepts, and translate studied content into daily life. This study revealed that the support of students with writing and reading skills increased academic performance during the formative assessments. It was also evident that activities prepared by teachers foster and increase reading skills especially when students are given the opportunities to interact while they are doing learning activities. This confirmed the demands of social constructivism theory, where much is expected and anticipated to the development of cognitive and critical thinking skills.

Further, students learn fairly through reading textbooks, which introduces them to the world of academic literacy and strengthens the understanding of subject concepts. Unfortunately, there is evidence that students are less engaged in reading textbooks in Tanzania secondary schools (Barrett & Bainton, 2016; Brock-Utne, 2014; Gabrieli et al., 2018). There is hope that the implementation of the language instructional objective activities introduced by this study into a lesson plan will help STEM teachers, particularly those teaching biology, to use potential short texts that would help students to improve English reading and listening skills.

Considering the relevance of bilingual instructions in teaching and learning, the questions evoked by this study are to know when and how the mentioned strategy can be used in classroom while teaching and learning STEM? In relation to biology subject, specifically invertebrate systematic, the relevance of the use of both English and Kiswahili was found very relevant. For example, using the binomial nomencla-ture, some biology teachers mentioned the name *Lumbricus terrestris* (Earthworm). However, in places where schools are located, it is difficult for a student to under-stand this binomial nomenclature because this name is far from the Kiswahili name

"Mnyoo wa ardhini" commonly used as fishing bait. For other teachers who started with Kiswahili name, students could remember and memorize the scientific name than the previous case. This is a sign that switching from English to Kiswahili is more helpful to clarify and provide an understanding of the subject content and its transfer into daily examples.

7 Conclusion

For students to learn STEM meaningfully, the teaching and learning need to acknowledge language pedagogical strategies which comprise of the use of language supportive activities, translate when necessary, interpret for students, help students to pronounce correctly, use of language genres specific to the subject and the topics, provision of a glossary and use of simple English sentences. As policy-practice implications, strategic use of Kiswahili and English together increased students' interactions and activeness in class which, in turn, improved the learning of content through English. As well, students viewed bilingual instructions as helpful to them and recommended that the approach be adopted in all classroom interactions.

8 Recommendations

For students to meaningfully learn STEM, this paper suggests that teaching and learning need to acknowledge language pedagogical strategies. Again, professional development can increase classroom writing and reading activities. Through training and workshops, STEM teachers can critically develop their ability in STEM teaching with language support. Again, it suffices to recommend the use of bilingual instructions to be articulated in Tanzania language in education policy.

Acknowledgements We owe a debt of gratitude to the Management of African Centre of Excellence for Innovative Teaching and Learning Mathematics and Science (ACEITLMS) based at the University of Rwanda, College of Education for funds and research permissions to conduct this study. We would also like to thank the management of St John's University of Tanzania for the assistance to obtain research permits in Tanzania. Further, we appreciate the contribution of the Language Supportive Teaching and Textbooks in Tanzania project (LSTT) team, particularly Dr. Angeline M. Barrett from the School of Education, University of Bristol, UK, for her constructive comments on this manuscript. Again, we are grateful to the participants for the breadth of information they shared with us.

References

Abdulaziz, M. (1982). Patterns of language acquisition and use in Kenya: Rural-urban differences. *International Journal of the Sociology of Language, 34*(1), 95–120.

Barrett, M. A., & Bainton, D. (2016). Re-interpreting relevant learning: An evaluative framework for secondary education in a global language. *Comparative Education, 52*(3), 392–407. https://doi.org/10.1080/03050068.2016.1185271.

Barrett, A., Mtana, N., Osaki, K., & Rubagumya, C. (2014). *Language supportive teaching and textbooks in Tanzania: Baseline study report.* Bristol: University of Bristol and Dodoma: University of Dodoma. Retrieved January 14, 2020, from http://www.educationinnovations.org/sites/default/files/LSTT%20baseline%20reportPSIPSE0pdf.

Brock-Utne, B. (2014). Language-in-education policies and practices in Africa with a special focus on Tanzania and South Africa. In J. Zadja (Ed.), *Second international handbook on globalisation, education and policy research* (pp. 615–631). Dordrecht: Springer.

Cooksey, B. (1986). Policy and practice in Tanzanian secondary education since 1967. *International Journal of Educational Development, 6*(3), 183–202. https://doi.org/10.1016/0738-0593(86)90016-7.

Coyle, D., Hood, P., & Marsh, D. (2010). *Content and language integrated learning.* Cambridge: Cambridge University Press.

Cummins, J. (2000). *Language, power, and pedagogy: Bilingual children in the crossfire.* Clevedon: Multilingual Matters Ltd.

Dalton-Puffer, C., & Smit, U. (2013). Content and language integrated learning: A research agenda. *Language Teaching, 46*(4), 545–559. https://doi.org/10.1017/S0261444813000256.

Davis, B., Sumara, D., & Luce-Kepley, R. (2000). *Engaging minds-learning and teaching in a complex world.* London: Lawrence Erlbaum Associates Publishers.

Gabrieli, P., Sane, E., & Alphonce, R. (2018). From access to quality secondary education: Developing language supportive textbooks to enhance teaching and learning of biology subject in Tanzania. *Journal of Education, Society and Behavioural Science, 25*(1), 1–15. https://doi.org/10.9734/JESBS/2018/40999.

Gahigi, M. (2008). *Rwanda: English language teaching kicks off*. AllAfrica.com. Retrieved January 11, 2020, from http://allafrica.com/stories/200812010940.html.

Galabawa, J. C. J. (2006). Implications of changing the language of instructions in secondary and tertiary education in Tanzania. In B. Brock-Utne, Z. Desai, & M. Qorro (Eds.), *Focus on fresh data on the language of instruction Debate in Tanzania and South Africa.* Dar es Salaam: African Minds. https://doi.org/10.1163/9789460912221011.

Knoch, U. (2011). Rating scales for diagnostic assessment of writing: What should they look like and where should the criteria come from? *Assessing Writing, 16*(1), 81–96. https://doi.org/10.1016/j.asw.2011.02.03.

Kothari, C. R. (2008). *Research methodology: Methods and techniques* (2nd ed.). New Delhi: New Age International Publishers.

Kyeyune, R. (2003). Challenges of using English as a medium of instruction in multilingual contexts: a view from Ugandan classrooms. *Language, Culture, and Curriculum, 16*(2), 173–184. https://doi.org/10.1080/07908310308666666.

Marsh, D. (Ed.). (2002). *CLIL/EMILE: The European dimension.* Jyväskylä: University of Jyväskylä.

Massler, U., Stotz, D., & Queisser, C. (2014). Assessment instruments for primary CLIL: The conceptualisation and evaluation of test tasks. *The Language Learning Journal, 42*(2), 137–150. https://doi.org/10.1080/09571736.2014.891371.

Maxwell, J. A. (2012). *Qualitative research design: An interactive approach* (Vol. 41). Sage.

MoEST. (2015). *Education and training policy.* Dar es Salaam: Ministry of Education, Science, and Technology.

MoEST. (2016). *Basic education statistics of Tanzania (BEST).* Dar es Salaam: Ministry of Education, Science, and Technology.

MoEST. (2017). *Basic education statistics of Tanzania (BEST)*. Dar es Salaam: Ministry of Education, Science, and Technology.

MoEST. (2018). *Basic education statistics of Tanzania (BEST)*. Dar es Salaam: Ministry of Education, Science, and Technology.

MoEST. (2019). *Basic education statistics of Tanzania (BEST)*. Dar es Salaam: Ministry of Education, Science, and Technology.

MoEVT. (2014). *Education and training policy*. Dar es Salaam: Ministry of Education and Vocational Training.

Mohan, B., & Slater, T. (2005). A functional perspective on the critical theory/practice relation in teaching language and science. *Linguistics and Education, 16*(2), 151–172. https://doi.org/10.1016/j.linged.2006.01.008.

Nyerere, J. (1968). *Education for self-reliance, freedom, and socialism: A selection from writings & speeches, 1965–1967*. Dar es Salaam: Oxford University Press.

Phillipson, R. (1992). *Linguistic imperialism*. Oxford University Press.

Puja, G. K. (2003). Kiswahili and higher education in Tanzania: Reflections based on a sociological from three Tanzanian university campuses. In Birgit Brock-Utne, Zubeida Desai, & Martha Qorro (Eds.), *Language of instruction in Tanzania and South Africa (LOITASA)*. E&D Limited: Dar-es-Salaam.

REB. (2015a). *Integanyanyigisho y'uburezi bw'inshuke kuva ku myaka 3 kugeza ku myaka 6*. Ministry of Education: Kigali.

REB. (2015b). *Competency based curriculum-summary of curriculum framework pre-primary to upper secondary*. Ministry of Education: Kigali.

REB. (2020). *Guidelines of using English language as medium of instruction*. Ministry of Education: Kigali.

Ricketts, A. (2014). Preservice elementary teachers' ideas about scientific practices. *Science & Education, 23*(10), 2119–2135.

Roy-Campbell, Z. (2001). *Empowerment through language*. Trenton, NJ and Asmara, Eritrea: Africa World Press.

Rubagumya, C. M. (1990). Language in Tanzania. In C. M. Rubagumya (Ed.), *Language in education in Africa: A Tanzanian perspective*. Philadelphia: Multilingual Matters.

Rubagumya, C. (2003). English medium primary schools in Tanzania: A new linguistic market in education? In B. Brock-Utne, Z. Desai, & M. Qorro (Eds.), *Language of instruction in Tanzania and South Africa (LOITASA)* (pp. 149–170). Dar es Salaam: E & D Limited. https://doi.org/10.1163/9789460912221005.

Samuelson, B. L., & Freedman, S. W. (2010). Language policy, multilingual education, and power in Rwanda. *Language Policy, 9*(3), 191–215. https://doi.org/10.1007/s10993-010-9170-7.

Shohamy, E., Gordon, C. M., & Kraemer, R. (1992). The effects of raters' background and training on the reliability of direct writing tests. *The modern Language Journal, 76*(1), 27–33. https://doi.org/10.2307/329895.

URT. (2019). *National Bureau of statistics in Tanzania: National Data*. Dodoma.

URT. (2020). *Ministry of education, science, and technology*. O-level Biology Syllabus for Secondary school. Form I-IV.

Vavrus, F. (2002). Postcoloniality and English: Exploring language policy and the politics of development in Tanzania. *TESOL Quarterly, 36*(3), 373–397.

Vygotsky, L. S. (1962). *Thought and language*. New York: Wiley.

Vygotsky, L. S. (1978). *Mind and society: The development of higher psychological processes*. Cambridge MA: Harvard University.

Wells, G. (1999). *Dialogic inquiry: Toward a sociocultural practice and theory of education*. Cambridge University Press.

Opanga David is currently a Ph.D. in Biology Education student at the African Centre of Excellence for Innovative Teaching and Learning Mathematics and Science (ACEITLMS), University of Rwanda, College of Education (UR-CE). David is also affiliated with St John's University of Tanzania where he serves as an Assistant Lecturer under the Faculty of Humanities and Education, and the Faculty of Natural and Applied sciences. He is currently involved in Biology Education projects, particularly inquiry-based learning, language supportive pedagogy, and study of invertebrate systematic.

Alphonse Uworwabayeho is a Senior Lecturer in education at the University of Rwanda-College of Education (UR-CE). He obtained his Ph.D. in Mathematics Education, specializing in the integration of ICT in the teaching and learning of mathematics. His research interest lies in teacher professional development on enhancing active teaching and learning. Currently, he is also leading the Department of Early Childhood and Primary Education. He is an associate member of the African Centre of Excellence for Innovative Teaching and Learning Mathematics and Science (ACEITLMS).

Théophile Nsengimana is a Ph.D. student in science education and Lecturer at the University of Rwanda, College of education. Nsengimana has a wide educational and research experience, resulting from local, regional, and international workshops and conferences. He is interested in Mathematics and science teacher education as well as science education. Nsengimana published different papers related to science curriculum implementation in internationally recognized peer-reviewed journals.

Evariste Minani graduated from the University of Cape Town (UCT) with a Ph.D. in physics. He is currently a physics lecturer at both undergraduate and postgraduate programs at the University of Rwanda-College of Education (UR-CE). He also completed a postgraduate certificate program in teaching and learning in higher education from UR-CE. Further, Minani is an associate member of the African Centre of Excellence for Innovative Teaching and Learning Mathematics and Science (ACEITLMS) based at UR-CE. Minani has published several papers in the area of materials science in different international peer-reviewed journals and currently his research focus is in Physics education.

Nsengimana Venuste has a Ph.D. in "Sciences Agronomiques et Ingénierie Biologique (Agronomy and Bio-engineering)" from the University of Liège—Gembloux Agro Bio-Tech, Belgium. In addition to the Bachelor's degree in Biology Education awarded by the former National University of Rwanda, Nsengimana has a Master's degree in Biodiversity Conservation and a Postgraduate Certificate in teaching and Learning in Higher Education awarded by the University of Rwanda. Nsengimana serves the University of Rwanda-College of Education (UR-CE) as a Lecturer of Biology. Further, Nsengimana is an associate member of the African Centre of Excellence for Innovative Teaching and Learning Mathematics and Science (ACEITLMS) based at the UR-CE. He is also an associate member of the Centre of Excellence in Biodiversity and Natural Resources Management (CoEB), based at the University of Rwanda-College of Science and Technology (UR-CST). His area of research is Biodiversity Conservation (Ecology and Conservation) and Biology Education.

Individual Language Planning for Self-Directed Learning in Multilingual Information Technology Classrooms

Jako Olivier

Abstract This chapter explores the affordances of individual language planning in terms of facilitating self-directed learning (SDL) in multilingual Information Technology (IT) classrooms at the high school level. Multilingualism is a reality within South African schools, however, not only does education in general in South Africa show evidence of the hegemony of English, in a subject such as IT, English is even more prominent. This chapter regards language as being one of the essential resources to be considered in terms of effective SDL in multilingual IT classrooms. The multilingual nature of classrooms and the status of English in South Africa pose unique challenges which have not been considered in terms of SDL in multilingual IT classrooms before and this serves as the impetus for this research. The problem posed for this chapter is, what affordances do individual language planning provide in terms of facilitating SDL in multilingual IT classrooms at the high school level? To this end, an exploratory qualitative study was undertaken regarding language practices within IT classrooms in high schools in the Free State province of South Africa. The aim of this chapter is to provide recommendations that can serve as practical guidelines for effective individual language planning in this context.

Keywords Individual language planning · Self-directed learning · Information technology · Language policy · Multilingual education

1 Introduction

The complex nature of multilingualism in the South African education context is clear (Heugh & Stroud, 2019). In this regard, South Africa has eleven official languages and many other minority languages and South Africans are very multilingual themselves (Charamba, 2020; Coetzee-Van Rooy, 2016). However, despite official recognition of eleven languages and historical bilingualism the language, English, dominates in

J. Olivier (✉)
Research Unit Self-Directed Learning, Faculty of Education, North-West University, Mahikeng, South Africa
e-mail: jako.olivier@nwu.ac.za

© The Author(s), under exclusive license to Springer Nature Switzerland AG 2021   117
A. A. Essien and A. Msimanga (eds.), *Multilingual Education Yearbook 2021*, Multilingual Education Yearbook,
https://doi.org/10.1007/978-3-030-72009-4_7

education at all levels. Olshtain and Nissim-Amitai (2004, p. 59) highlight the importance of autonomy on the side of learners in terms of language use. Consequently, the concept of individual language planning is relevant as individuals also engage in such language planning activities (Orman, 2008; Zhao & Baldauf, 2012). Despite the logical association of student individual language planning with learners, the focus in this chapter is more on teachers and the classroom. Furthermore, the emphasis is on how a context conducive of such language planning can be created in terms of self-directed learning (SDL) within the Information Technology (IT) classroom context.

Within this context, the SDL of the subject IT (Department of Basic Education, 2011; Mentz et al., 2012) is the aim. SDL implies that learners take charge of their own learning (Brockett & Hiemstra, 2019; Bosch et al., 2019; Francom, 2010) and it is proposed that individual language planning is considered a part of this process. To this end, the role of language in terms of technology is considered.

The importance of language in terms of the integration of technology in the classroom, which is essential for IT teaching and learning, is evident from the literature. In this regard, Gudmundsdóttir (2010) acknowledges the relevance of recognizing cultural as well as linguistic aspects and specifically an appropriate language of instruction in terms of computers in education. Furthermore, Adams (1998) emphasizes that computer-related interactions are in fact language interactions in a broader perspective, therefore, this issue becomes very complex in a multilingual context. This study, however, focuses specifically on the subject IT as is taught in South African high schools.

In terms of the teaching and learning of IT and a similar subject such as Computer Applications Technology a number of challenges are raised in the literature (Olivier, 2011a; Barlow-Jones et al., 2014; Fambaza, 2012; Havenga & Mentz, 2009; Mentz et al., 2012; Venter, 2016). In this regard, Mentz et al. (2012) noted issues around sufficient hardware and software as well as access to computers outside of school in reference to research in two rural schools in the North West Province. However, in this study, the issue of language was not raised. Consequently, this research aims to address the gap in the literature regarding the intersections between individual language planning, SDL and multilingual Information Technology (IT) teaching.

This chapter draws on exploratory research where data gathered from high school teachers by means of questionnaires and interviews (cf. Olivier, 2011a) with subject experts as well as observations conducted in two classes. In addition, a theoretical framework regarding relevant recent publications is provided. From this data an overview of the current situation is provided, and a set of guidelines are proposed towards effective individual language planning for SDL in multilingual IT classrooms. This chapter reports on research related to a study of which parts have been reported in Olivier (2011a, 2011b, 2013). However, in this chapter, the focus has shifted from multilingualism to individual language planning specifically. The following research question is posed for this study: What affordances do individual language planning provide in terms of facilitating SDL in multilingual Information Technology (IT) classrooms at the high school level?

2 Literature Review

2.1 Individual Language Planning and Multilingual Education

The concept of individual language planning should be considered in terms of the scholarship around language planning. Language planning is generally considered in terms of it being realized as status planning (which relates to the position of a language in society), corpus planning (regulating the internal language structure in terms of spelling for example), language-in-education planning (emphasizing the role of language for and as an object of learning) as well as prestige planning (addressing the image of a language) (Olivier, 2011a; Fishman, 1974; Zhao & Baldauf, 2012). The distinction between macro, meso and micro language planning is also used widely in the language planning literature. Notably, Fishman (1974) describes language planning practices as "decision making in connection with language problems" (p. 15). The focus, however, in this chapter is on individual language planning and hence this research relates to solving language problems at the level of the individual.

As stated before, the focus in this chapter is on the individual. Zhao and Baldauf (2012) highlight the importance of individual agency in terms of language planning activities. Furthermore, Zhao and Baldauf (2012) distinguish between the following three types of people who they consider show individual agency in terms of language policy and planning: "People with expertise", "People with influence" and "People with power" (p. 6). Yet, it is proposed here that all language users are involved in a language planning process albeit not formal or overt. Not only is language unique for each individual (Günther, 2016), individuals make daily choices in selecting different language varieties and languages to communicate and ultimately learn. These choices can also be described as a process of individual language planning.

However, the concept of individual language planning should not just be regarded within the language practices within a wider community and country but also specifically smaller social units like the family. In this regard, family language policy (cf. Fogle, 2013) is defined by Curdt-Christiansen (2009, p. 352) as "a deliberate attempt at practicing a particular language use pattern and particular literacy practices within home domains and among family members". The language of an individual is part of their identity and in this context, Wright (2016) observes that language can be considered a strong marker of identity. In South African classrooms, such identities come into play as learners negotiate not only their place in the classroom but also as they strategically plan the use of language for the sake of socialization and learning.

This process of taking charge of your language practices, either under the influence of others or independently, may provide opportunities to support a sense of agency. Zhao and Baldauf (2012) acknowledge that "individual agency at the micro level, which often reveals how language policies trickle down to the local communities and contexts, has an important place in the process of building up prestige" (p. 18). In this regard, Liddicoat and Baldauf (2008) observe that at a micro-level language planning is conducted by individuals or smaller groups who can promote or even revive languages. This is the level of language planning at play in this chapter.

The definition of a mother tongue by Skutnabb-Kangas (1988) where this concept is regarded in terms of a person's origin, competence, function and identification supports the view of individual language planning. Importantly, when it comes to identification, this is done by speakers themselves internally or externally by association with others. In addition, even in terms of language rights this is regarded as individual rights (Wright, 2016). The conceptualization of individual language planning in this chapter also relates not to the work of an individual towards status planning of a language for the sake of the wider language community but rather the practices and choices made by individuals in terms of their language selection and use: their *autonomoglottal praxis* as it were.

Individual language planning can take place by means of different strategies ranging from a mere selection of a single language or the mixing of languages and even varieties within utterances and other instances of language. One such strategy is code-switching. According to Keller (2020) classic code-switching involves utilizing two or more languages within a conversation and does not imply insufficient knowledge of either language, but rather strategic use of available lingual resources. For Gardner-Chloros (2009) code-switching is "the use of several languages or dialects in the same conversation or sentence by bilingual people" (p. 4). Furthermore, the use of code-switching in the classroom setting has been widely researched (Gardner-Chloros, 2009). This aspect also prompts the need to consider translanguaging.

From the literature, it is apparent that the borders between languages and codes are not so clear in multilingual contexts and quite often in such circumstances translanguaging is common (Duarte, 2020; Makalela, 2013). Translanguaging is defined by García (2009, p. 140) as an "act performed by bilinguals of accessing different linguistic features or various modes of what are described as autonomous languages, in order to maximize communicative potential" this process includes code-switching. Furthermore, the concept of translanguaging is extended to *multilanguaging* or *ubuntu translanguaging* by Makalela (2018) who describes this as "instances where speakers have acquired more than two languages simultaneously and where there is more than one language of input and output in a discourse for meaning making" (p. 5). Yet, Duarte (2020) also notes criticism against translanguaging as "translanguaging pedagogy clashes against prevailing monolingual ideologies often translated into immersion models for language teaching which lead to strict language separation" (p. 234). But the research conducted by Duarte showed that translanguaging practices fostered dynamic plurilingual practices. However, for this chapter, the specific South African context needs to be considered.

In the South African context, language and language planning should be considered in terms of historical colonial bilingual policies, followed by *de jure* multilingualism and de facto monolingual English hegemony (Mathole, 2016; Olivier, 2011a). However, the wider South African and school populations are very multilingual (Ndebele, 2014; Olivier, 2011a). In this chapter, language is also regarded as an important resource in the process of SDL.

2.2 Self-Directed Learning (SDL)

Central to the way the concept of SDL is approached in this chapter is the classical definition of this concept as formulated by Knowles (1975, p. 18) where it is considered "a process in which individuals take the initiative, with or without the help of others, in diagnosing their learning needs, formulating learning goals, identifying human and material resources for learning, choosing and implementing appropriate learning strategies and evaluating learning outcomes". In considering the definition of SDL, this chapter proposes that language should also be considered as an essential resource in the learning context. In this regard, the choice of language for learning is a conscious act by the learner which could be in interaction with others or in terms of resources. The description of language as a "resource" is also common in language planning literature (Catalano & Hamann, 2016).

SDL places the learner at the core of the learning process. As is evident from the definition by Knowles above, the formulation and reaching of goals set by learners guides this process. However, from the nature of curricula, the manner in which schooling is structured and prevailing practices of teachers, goals are often determined on behalf of learners. Hence, the following remark by Johnson and Johnson (2019) is highly relevant: "Much of SDL begins with teacher-directed learning that assumes that gradually through dialogue and discussions with the teacher and classmates, students will internalize the goals and the responsibility of learning will shift to the students" (p. 42). Therefore, the teacher plays a key role in the path towards SDL as the process needs to be supported and facilitated (cf. Francom, 2010).

SDL is not only a learning process it is also a learner characteristic. Brockett and Hiemstra (2019) note that "self-direction in learning refers to both the external characteristics of an instructional process and the internal characteristics of the learner, where the individual assumes primary responsibility for a learning experience" (p. 56).

Despite the extensive literature on SDL and andragogy, SDL is highly relevant in the school context (cf. Van Deur, 2017) and not only adult education (Brockett & Hiemstra, 2019). Despite early associations of SDL with adult education (Knowles, 1975) this concept is increasingly also considered in the school context (Brockett & Hiemstra, 2019). Hence, this concept is relevant for this research.

Different learning strategies can be employed in order to foster self-directedness among learners and these include cooperative learning (cf. Bosch et al., 2019; Johnson & Johnson, 2019; Lubbe, 2020), process-oriented learning (Bosch et al., 2019) and problem-based learning (Bosch et al., 2019). Cooperative learning relates to "students working together to maximize their own and each other's learning" (Johnson & Johnson, 2019, p. 38) and is highly relevant in the context of SDL. According to Bosch et al. (2019), learning through association with other individuals in knowledge production is essential. According to Johnson and Johnson (2019) cooperative learning requires positive interdependence, a form of individual accountability, promotive interaction in the classroom, certain social skills as well as a measure of group processing. These aspects rely on language to ensure the interactions between learners are effective. In addition, on a practical level in the creation of groups for the sake of cooperative learning, learners should ideally be given the option to form groups themselves, for example, Lubbe (2020) found that respondents opted to form their own groups for the sake of language.

Furthermore, certain skills have been identified as being needed for self-directed learners. Lubbe (2020) provides a summary of SDL skills based on key SDL sources. For a number of these skills language is essential in order to obtain, process or share information or facilitate interactions. In addition, Francom (2010) distinguishes between four principles to enable the fostering of SDL skills:

- match the level of SDL required in learning activities to student readiness;
- progress from teacher to student direction of learning over time;
- support the acquisition of subject matter knowledge and SDL skills together; and
- have students practice SDL in the context of learning tasks.

These principles can also be considered in any attempt to facilitate individual language planning for SDL. Furthermore, the issue of multilingual learning in IT classrooms needs to be considered.

2.3 Multilingual Learning in Information Technology (IT)

The multilingual nature of South African schools necessitates the accommodation of the various languages used by learners. The affordances of technology in terms of facilitating multilingual education have been evident in subjects such as mathematics (Chikiwa & Schäfer, 2016; Libbrecht & Goosen, 2016) and natural sciences (Charamba, 2020). In this chapter, the focus is on language use in the subject IT. This subject is described in the subject's Curriculum and Assessment Policy Statement (CAPS) as follows:

> Information Technology is the study of the various interrelated physical and non-physical technologies used for the capturing of data, the processing of data into useful information and the management, presentation and dissemination of data. Information Technology studies the activities that deal with the solution of problems through logical and computational thinking. It includes the physical and non-physical components for the electronic transmission, access, and manipulation of data and information. (Department of Basic Education, 2011, p. 8).

Apart from theoretical content—available only in English and Afrikaans textbooks—this subject also has a practical component based around computer programming. In this regard, another level of language interaction is present as learners are expected to code in a specific computer language (cf. Goosen et al., 2007). However, there has been some work done in terms of the inclusion of African languages in this and similar contexts.

There has also been a number of studies where the affordances of using African languages in the classroom at school and university level in the context of using information communications technology has been evident (Olivier, 2013; Dalvit et al., 2005; Ndebele, 2014; Njobe, 2007). According to Libbrecht and Goosen (2016) "[f]or multilingual learners, the ability of a software application to speak multiple languages offers them a flexibility that may support them" (p. 225). In terms of South African IT classrooms, Olivier (2011b) observes that "[t]he subject IT is generally associated with English due to the subject content" (p. 217).

The discussed practices in IT classrooms should be considered in terms of multilingual education. Duarte (2020) describes multilingual education as "an umbrella-term for various school approaches including several languages of instruction, also for those aiming at fostering elite bilingualism" (p. 233). García and Lin (2017) state that "[t]he challenge for schools in the twenty-first century is how to create flexible dynamic models of bilingual education, where students' language practices are used not simply as a 'scaffold' when learning in a second language, but as a transformative practice that puts power back in the lips of multilingual speakers instead of simply acquiescing to the power of education and state authorities" (p. 17). Consequently, this chapter subscribes to the fostering of such transformative practices through individual language planning.

Within the aforementioned context, the issue of epistemic distance in terms of language is relevant. In this chapter, epistemic distance is interpreted in the same manner as the way the term "epistemic" is conceptualized with regard to epistemological access (cf. Charamba, 2020). For Morrow (2007) epistemological access relates to "access to knowledge" and teaching is "the practice of enabling epistemological access" (p. 2). The term *epistemic* is used here in the Greek sense of ἐπιστήμη or epistēˊmē which relates to knowledge. It is proposed in this chapter that *epistemic distance* relates to the degree of lack of knowledge due to "situational barriers" (Brockett & Hiemstra, 2019, p. 274). In this research, epistemic distance can also be due to the language distance occurring due to different languages used by teachers, learners and in content. It is, however, important to note that both epistemic and language distance imply a continuum and that they may involve highly complex individualized contexts.

The next section provides an overview of the research methodology of this research in which the affordances of individual language planning in terms of facilitating SDL in multilingual IT classrooms at the high school level were explored.

3 Methodology

3.1 Research Design

The data presented in this chapter are part of a wider mixed-method study (Olivier, 2011a) of which the qualitative data are revisited in terms of individual language planning and SDL specifically. To this end, this basic qualitative research (Merriam, 2009) followed an interpretivist paradigm (Bakkabulindi, 2015) in order to probe the lived experiences within the IT classrooms context.

3.2 Sampling

This research involved teachers, subject experts and two schools. For this research purposive sampling was employed as the research participants were chosen "on the basis of their judgement of their typicality or possession of the particular characteristic(s) being sought" (Cohen et al., 2017, p. 474). Similarly, the two schools were also chosen purposively as they represented two typical types of schools where IT is presented in the Free State.

The first group of research participants included teachers and the inclusion criteria for this group involved that they had to be IT teachers in the Free State province teaching grade 10. Here, 11 of the 14 teachers—this covered all IT teachers in the Free State province who taught grade 10s at the time of the data collection—that adhered to the inclusion criteria responded to the invitation to take part in the research. In addition, two schools were selected where observations were made in grade 10 IT classes. The second group of research participants were four subject experts involved in IT and technology at the provincial and national level.

Finally, two IT classes at two different schools were chosen by means of convenience sampling for observations. The two schools used for the observations in this research were a multilingual highly resourced school in a town with mainly Afrikaans and English-speaking teachers, called School A, as well as another, a township school from a disadvantaged background with teachers who speak various African languages, called School B.

3.3 Data Collection

The part of this study reported here relates to a questionnaire conducted with IT teachers in the Free State province, interviews with subject experts and observations in classrooms in two selected schools.

The first part of the research involved two sets of participants and methods of data collection: questionnaires completed by teachers and interviews with subject experts (cf. Olivier, 2011a). The structured questionnaire used for this research involved both open-ended and closed-ended questions probing among other things the language practices within IT classrooms in the province. This part of the research was quite quantitative in nature as the closed questions allowed for the recording of predetermined choices and numeric data while in terms of qualitative data the open-ended questions allowed the research to gain insight into the teachers' opinions and perceptions. The response rate of the IT teachers in the Free State province was 63.5% and it was evident that a representative sample was reached as the sample showed sufficient heterogeneity in terms of the IT teachers in the province. However, only the questions relevant to the focus of this discussion are reported in this chapter.

Furthermore, data was also collected through observations at the two schools noted in the previous section. The data reported here emanates from field notes and recordings made.

3.4 Research Ethics

This research adhered to the ethical guidelines as set out by the institution that provided ethical clearance as well as the requirements of the provincial department of education. Hence, an ethics application was made and ethical clearance granted by a research ethics committee at the university from which this research was done. Permission was granted by the Free State Department of Education for the research to take place. The researcher ensured fair recruitment and informed consent practices. Where learners were involved, parental permission and learner assent were obtained. Furthermore, confidentiality and privacy were ensured throughout the process and data were stored securely. All learners consented to be recorded and parents provided permission for this to take place. Measures were taken to ensure the confidentiality and privacy of teachers and learners involved in the recordings specifically.

3.5 Data Analysis

The qualitative data were analysed inductively through the identification of codes and themes from the interviews, open-ended questions and observations (Saldaña, 2011). Some descriptive statistics relevant to this chapter are presented below.

3.6 Main Findings

3.6.1 Questionnaires

The data from the questionnaires provide an in-depth snapshot of the nature of language in the researched IT classrooms in the Free State. More details on these aspects have already been reported (cf. Olivier, 2011b). At the time of this study, the majority of IT teachers in the Free State had Afrikaans as a mother tongue followed by speakers of Sesotho and Setswana while none of the IT teachers in the Free State had English as a mother tongue. This is quite significant as English is generally used as the language of learning and teaching. However, it was clear that apart from most teachers at least being bilingual the teachers also knew other languages such as isiZulu and isiXhosa in addition to the languages mentioned above. The prominence of Afrikaans and Sesotho is in line with the language demographics of the province.

In terms of the language profile of learners, it was noted that in 91% of the schools learners spoke more than one language. Apart from the one school where only Afrikaans was used, the rest of the schools were very multilingual. The most prominent mother tongue among learners was Sesotho ($n = 235$; 63.5%), followed by Afrikaans ($n = 63$; 17%), English ($n = 33$; 8.9%), isiZulu ($n = 25$; 6.8%), Setswana ($n = 7$; 1.9%); isiXhosa ($n = 3$; 0.8%); and Xitsonga ($n = 2$; 0.5%) as well as one learner speaking Greek and another Hindi. In terms of the formal language of learning and teaching of IT English is therefore also very prominent in this province. Hence, the use of other languages was an attempt at conscious language planning and was usually done at a communicative or supportive level.

The teachers also indicated on the questionnaire how they would accommodate the language diversity in the classes and it was evident that the most common way in which multilingualism was accommodated and promoted was through code-switching, using textbooks in different languages (Afrikaans and English textbooks are used) as well as by means of terminology lists. However, it is clear that most teachers do not actively use or promote the use of other languages in these classes.

3.6.2 Interviews

Some interesting perspectives were presented by the experts in terms of language in the IT classroom context. Firstly, the prominence of English was clear. In this regard respondent 1 noted that "English is the preferred medium of use at this point in time", but also that the respondent believed that "languages have a place in the curriculum, but it might be difficult to offer all subjects in all the official languages in the country". Respondent 2 made the following remark in terms of accommodating multilingualism "Possibly in the future and would be recommended—English at present predominates both as a language of learning and on the Internet". Respondent 3 also noted that little is done formally in terms of accommodating languages other than Afrikaans and English.

In further discussions, the nature of the subject in terms of language was also noted. Respondent 3 stated that "programming language can be seen as independent but all the coding words that are used in about all the programming languages are mostly English" and hence there is potentially a significant distance between the language used by learners and that used for teaching and ultimately doing computer programming itself. The nature of the programming language as being based on English prompts the first level of both language and epistemic distance.

General issues around language in education were also noted. Respondent 4 observed that "In terms of the learners' language command, it could have a big effect. Language is the basis of all learning and if a learner does not have a good command of the language of learning and teaching, it definitely affects understanding and comprehension and will have an effect on the learning and teaching".

The promise of individual language practices allowing for the inclusion of additional languages in the classroom context shows promise as Respondent 3 admitted that through the use of additional languages "learners with a language barrier can benefit a lot because technological facts and concepts, as well as most of the problem-solving effects and concepts used when programming is been taught".

3.6.3 Observations

In the lessons observed two distinct trends emerged. In School A, the interactions were either in Afrikaans and English with the majority of teacher-related interactions as well as electronic and paper-based content being in English. This was despite the fact that the learner population was very multilingual. The learning was very teacher-centred at both schools and this also had an effect on the nature of learner participation and the creation of an active and collaborative space.

From the observations, the ongoing prominence of English was also very evident in both schools and this is in line with other observations in South African classrooms (Olivier, 2011b; Gudmundsdóttir, 2010). The nature of IT learning content and especially the programming context relies heavily on English vocabulary. Hence, this context would require sufficient English knowledge with other languages acting only in a supportive capacity. However, the drive to use English ties in with this language's status as a vehicle towards social mobility. Interestingly, even in School A there was a trend for Afrikaans-speaking learners to opt to also function mainly in English in interactions with peers, the teacher and content. Generally, however, learners used their mother tongues in some classroom discussions. Yet, it is clear that more can be done to encourage the use of other languages in this context. In this regard, it is evident that depending on learners' language there seems to be varying degrees of epistemic distance between the language of learning and teaching and especially the programming language. It is, therefore, evident that due to the nature of programming languages and the prevalence of English as language of learning and teaching, the mother tongue of learners would determine the epistemic distance between learner and learning content. In this regard, there seems to be a continuum in terms of distance ranging from English, Afrikaans (which is related linguistically to English), languages spoken by a number of learners in a class (such as Sesotho) and then isolated languages (such as isiZulu in this case).

In both schools, elements of translanguaging were present as learners interacted with each other. Learners were often grouped together in terms of languages they shared and would work collaboratively in their mother tongues. Such acts would close the epistemic distance in terms of language, but also isolated learners linguistically in some cases. Both teachers also employed translanguaging strategies. In School A, general instructions were provided in English with a shortened summary in Afrikaans. The lesson content was shared mainly in English. But most of the interactions were in English unless in one-to-one situations between teacher and learner. In School B, instructions were provided in Sesotho with limited code-switching to English. When it came to the sharing of lesson content then this was done in English. With the introduction of content in Sesotho for example, during the intervention (Olivier,

2011a), learners reacted positively and this prompted increased usage of Sesotho. In the next section, the analysed data is discussed in terms of the set research question.

4 Discussion

At the start of this chapter, the aim was set to explore the affordances that individual language planning provides in terms of facilitating SDL in multilingual Information Technology (IT) classrooms at the high school level. It is evident that a degree of individual language planning is already taking place and that through developing teacher skills, content and learning activities in terms of language as a resource clear benefits are evident. However, it is essential that teachers are also aware of a dynamic language and epistemic distances in this regard.

From the data portrayed in this chapter, it is clear that IT classrooms in the Free State tend to be focused on English despite existing multilingualism. These aspects are evident from the language profiles found as well as the descriptions and observations of practices in this research context. From the responses by subject experts and teachers, there are possibilities in terms of multilingualism but there is also a definite need for teachers to be empowered in terms of their own language capabilities and fostering individual language planning to create a context that is open to more languages. From the observations the important role of translanguaging as praxis of individual language planning is evident. However, the contextual factors relating to epistemic and language distance seems to be important and should be negotiated by teachers in order to create conducive circumstances for self-directed individual language planning.

In this chapter, there needs to be a distinction between language distance and epistemic distance. Language distance is the number of languages a learner is removed in terms of his or her mother tongue in comparison with the language of learning and teaching or even a programming language. Yet, in certain circumstances, the epistemic distance can be reduced by introducing additional languages in the learning context.

It is proposed in this chapter that the language distance is reduced in terms of facilitating the use of different languages in the classroom as well as making resources available in different languages. The situation can be summarized diagrammatically in Fig. 1.

Figure 1 depicts that only English-speaking learners (Scenario 1) in an English class have opportunities in terms of epistemic lingual access to a programming language based on English. Hence, the language distance is less. The language distance is also not so far for Afrikaans-speaking learners (Scenario 2) who might have access to Afrikaans books and due to the fact that the language is closer to English. However, when it comes to a language like Sesotho where there might be other learners who share the language in the class (Scenario 3) the language distance between the learner's mother tongue and the language used for learner-to-learner interactions are short but not necessarily in terms of the teachers and definitely not in

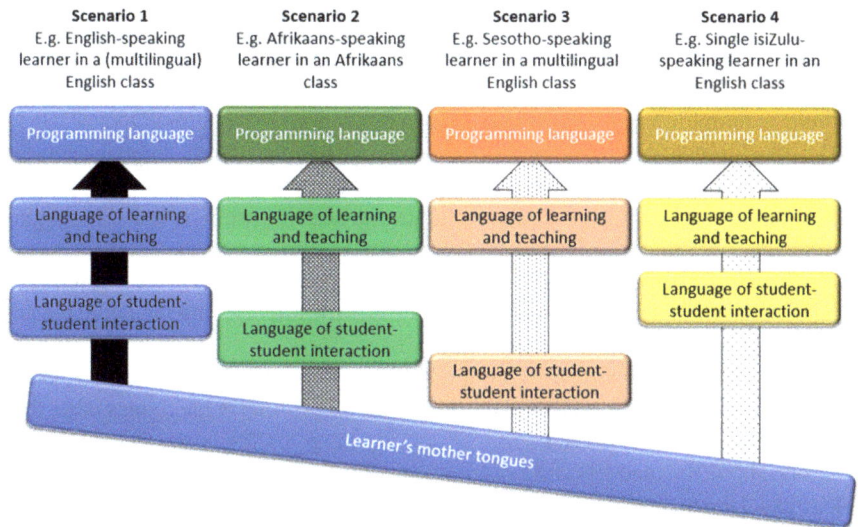

Fig. 1 Language and epistemic distances in IT classrooms

terms of the programming language. Finally, the instance with the furthest language distance between learner language and peer interaction, teacher and programming language (Scenario 4) relates to examples where a learner's mother tongue is only spoken by that learner and no other peers. The figure shows despite some overlap in language and epistemic distance, these two concepts are still different. The level of understanding or epistemic access is reflected in the way the arrows are filled in. Consequently, the continuum of epistemic distance between the identified four possible learner categories is evident based on the relevant language distance. Individual language planning efforts could shorten some of these distances, but that would require multilingual resources and multilingual teachers and peers.

Despite some observed success in terms of the introduction of content in languages other than English in the IT classroom, it is clear that for wider accommodation of multilingualism and ultimately the support of individual language planning certain external perception and language-specific issues need to be resolved. Ndebele (2014) also highlights some of these issues: insufficient discipline-specific and scientific terminology, negative language attitudes, weak bi- and multilingualism and orthographic inconsistencies. Such issues can, however, be countered by planned status and corpus planning and even language attitude planning (Verhoef, 1998).

Towards individual language planning for the facilitation of SDL in multilingual IT classrooms the following recommendations are made:

- Towards individual language planning teachers and learners should be encouraged to embrace multilingualism and exploit their knowledge of different languages in the classroom in order to close the language and epistemic distance.
- Schools should build on the existing language expertise of teachers and learners.

- Multilingual resources should be developed by experts in conjunction with IT teachers who have knowledge not only about their teaching contexts but also the languages and varieties appropriate to their learners. Such multilingual resources could potentially be shared online as open educational resources.
- The development of the mentioned resources can become handy language sources to aid terminology development and standardization.
- Teachers should be empowered to facilitate and actively promote translanguaging strategies.
- Teachers need to be informed on how to foster SDL especially also in terms of supporting individual language planning efforts towards closing the language and epistemic distance between them and learners.
- Aspects of IT classes that lend them to the use of mother tongues, such as group discussion, problem-solving and the writing of algorithms in languages other than English should be promoted.
- Efforts should be made to avoid stigmatization regardless of the language choice of learners. Hence, teachers should be made aware of the language needs of learners, their parents and the community.
- Strategies supporting SDL such as elements problem-based learning and cooperative learning should be structured around activities allowing for the use of different languages and supporting individual language planning practices such as the observed translanguaging.

5   Conclusion

This chapter explored the affordances of individual language planning for the facilitation of SDL in multilingual IT classrooms at the high school level. This premise builds upon the idea that languages should be considered as essential resources in the classroom setting. The specific content of IT as a subject leads to a greater influence of English than in other subjects. However, for effective SDL if learners are empowered to be able to select and effectively use other languages through translanguaging as resources.

As with other subjects, in IT further infusion of different languages are essential to counter what Gudmundsdóttir (2010) describes as follows: "Without an emphasis on learners' home language in school, learners can risk falling into a double literacy trap when they are expected to both learn a subject and take on new skills through the medium of an unfamiliar language" (p. 186). This was also evidenced from the analysis of the data in this chapter.

An important recommendation for IT classrooms remains that languages other than English be used actively. This concurs with the findings by Njobe (2007) where respondents in his study were of the opinion that "the language problem could be overcome by introducing isiZulu as the medium of instruction in the IT learning environment" (p. 2).

In order to facilitate cooperative learning, learners need to be able to communicate effectively and this may imply using languages other than the language of learning and teaching. In addition, this might also imply learners employing code-switching or translanguaging strategies in order to interact, formulate goals and create conditions to reach such goals.

Ultimately, if teachers create circumstances conducive and supportive towards the accommodation of multilingualism then learners would be able to enact individual language planning for SDL. In terms of the process of reaching goals within the context of SDL and cooperative learning, Johnson and Johnson (2019, p. 53) acknowledge "students often need both academic support to help them reach the goal and personal support to encourage them to persist and keep trying". This is also true for the process of facilitating individual language planning.

This study was very limited as it only focussed on a single province and selected schools. Therefore, more research is necessary to probe the mentioned aspects in different settings whether urban or rural. In addition, different language classroom demographics might also have an additional effect on the nature of individual language planning that is possible or needed.

Towards the realization of individual language planning for SDL it is proposed that learners are empowered to draw on different languages in the process of taking initiative to learn something. This can happen through interaction and negotiation with others in languages and varieties appropriate to the individuals involved which in turn would allow for effective diagnosis of learning needs and the formulation of learning goals in understandable language. Learners can also then identify not only human and material resources but also, importantly, resources in languages or codes relevant to their needs. Finally, based on the aforementioned appropriate learning strategies can be chosen and implemented and ultimately the learning outcomes be evaluated.

In conclusion, if Seemiller and Grace (2019), in the context of Generation Z claim that "coding is the new cursive", then it would be the ideal that such coding is done through the medium of learners' mother tongues.

References

Adams, A. (1998). Language awareness and information technology. In W. Tulasiewicz & J. Zajda (Eds.), *Language awareness in the curriculum* (pp. 41–50). Albert Park: James Nicholas.

Bakkabulindi, F. E. K. (2015). Positivism and interpretivism: Distinguishing characteristics, criteria and methodology. In C. Okeke & M. van Wyk (Eds.), *Educational research: An African approach* (pp. 19–38). Oxford: Oxford University Press.

Barlow-Jones, G., Van der Westhuizen, D., & Coetzee, C. (2014). An investigation into the performance of first year programming students in relation to their grade 12 computer subject results. In J. Viteli & M. Leikomaa (Eds.), *Proceedings of EdMedia 2014–world conference on educational media and technology* (pp. 77–83). Tampere, Finland: Association for the advancement of computing in education (AACE). Retrieved July 29, 2020 from https://www.learntechlib.org/primary/p/147485/.

Bosch, C., Mentz, E., & Goede, R. 2019. Self-directed learning: A conceptual overview. In E. Mentz, J. de Beer & R. Bailey (Eds.), *Self-Directed Learning for the 21st Century. Implications for Higher Education*. NWU Self-Directed Learning Series Volume 1 (pp. 1–36). Cape Town: AOSIS.

Brockett, R. G., & Hiemstra, R. (2019). *Self-direction in adult learning: Perspectives on theory, research, and practice*. London: Routledge. https://play.google.com/books/reader?id=-LF5DwA AQBAJ.

Catalano, T., & Hamann, E. T. (2016). Multilingual pedagogies and pre-service teachers: Implementing "language as a resource" orientations in teacher education programs. *Bilingual Research Journal, 39*(3–4), 263–278.

Charamba, E. (2020). Translanguaging in a multilingual class: A study of the relation between students' languages and epistemological access in science. *International Journal of Science Education*, 1–20. https://doi.org/10.1080/09500693.2020.1783019.

Chikiwa, C., & Schäfer, M. (2016). Teacher code switching consistency and precision in a multilingual mathematics classroom. *African Journal of Research in Mathematics, Science and Technology Education, 20*(3), 244–255.

Coetzee-Van Rooy, S. (2016). Multilingualism and social cohesion: Insights from South African students (1998, 2010, 2015). *International Journal of the Sociology of Language, 2016*(242), 239–265.

Cohen, L., Manion, L., & Morrison, K. (2017). *Research Methods in Education* (8th ed.). London: Routledge.

Curdt-Christiansen, X. L. (2009). Invisible and visible language planning: Ideological factors in the family language policy of Chinese immigrant families in Quebec. *Language Policy, 8*(4), 351–375.

Dalvit, L., Murray, S., Mini, B., Terzoli, A., & Zhao, X. (2005). Computers and African languages in education: An ICT tool for the promotion of multilingualism at a South African university. *Perspectives in Education, 23*(4), 123–131.

Department of Basic Education. (2011). *Curriculum and assessment policy statement grades 10-12: Information technology*. Pretoria: Government Printing Works.

Duarte, J. (2020). Translanguaging in the context of mainstream multilingual education. *International Journal of Multilingualism, 17*(2), 232–247.

Fambaza, T. (2012). *The experiences of teachers about teaching computer applications technology at FET band. (Unpublished master's dissertation)*. Durban: University of KwaZulu-Natal.

Fishman, J. A. (1974). Language planning and language planning research: The state of the art. In J. A. Fishman (Ed.), *Advances in language planning* (pp. 15–33). The Hague: Mouton.

Fogle, L. W. (2013). Parental ethnotheories and family language policy in transnational adoptive families. *Language Policy, 12*(1), 83–102.

Francom, G. M. (2010). Teach me how to learn: Principles for fostering students' self-directed learning skills. *International Journal of Self-Directed Learning, 7*(1), 29–44.

García, O. (2009). Education, multilingualism and translanguaging in the 21st century. In A. K. Mohanty, M. Panda, R. Phillipson, & T. Skutnabb-Kangas (Eds.), *Multilingual education for social justice: Globalising the local* (pp. 140–158). Bristol: Multilingual Matters.

García, O., & Lin, A. (2017). Extending understandings of bilingual and multilingual education. In O. García, A. Lin, & S. May (Eds.), *Bilingual and multilingual education* (pp. 1–20). Cham: Springer.

Gardner-Chloros, P. (2009). *Code-switching*. Cambridge: Cambridge University Press.

Goosen, G. L., Mentz, E., & Nieuwoudt, H. (2007). Choosing the "best" programming language. In *Proceedings of the computer science and IT education conference* (pp. 269–282).

Gudmundsdóttir, G. B. (2010). When does ICT support education in South Africa? The importance of teachers' capabilities and the relevance of language. *Information Technology for Development, 16*(3), 174–190.

Günther, F. (2016). *Constructions in cognitive contexts: Why individuals matter in linguistic relativity research*. Berlin: Walter de Gruyter.

Havenga, M., & Mentz, E. (2009). *The school subject information technology: A South African perspective.* In Proceedings of the 2009 Annual Conference of the Southern African Computer Lecturers' Association (pp. 76–80).

Heugh, K. & Stroud, C. (2019). Multilingualism in South African education: A southern perspective. In R. Hickey (Ed.) *English in Multilingual South Africa: The Linguistics of Contact and Change.* Studies in english language (pp. 216–238). Cambridge: Cambridge University Press.

Johnson, D. W. & Johnson, R. T. (2019). The impact of cooperative learning on self-directed learning. In E. Mentz, J. de Beer & R. Bailey (Eds.) *Self-Directed learning for the 21st century. Implications for higher education.* NWU Self-Directed learning series volume 1. (pp. 37–66).Cape Town: AOSIS.

Keller, M. L. (2020). *Code-Switching: unifying contemporary and historical perspectives.* Cham: Palgrave Macmillan.

Knowles, M. S. (1975). *Self-directed learning: A guide for learners and teachers.* Chicago, IL: Follett.

Libbrecht, P., & Goosen, L. (2016). Using ICTs to facilitate multilingual mathematics teaching and learning. In R. Barwell et al. (Eds.), *Mathematics education and language diversity.* New ICMI study series (pp. 217–235) Springer: Cham.

Liddicoat, A. J., & Baldauf, R. B. (2008). Language planning in local contexts: Agents, contexts and interactions. In A. J. Liddicoat & R. B. Baldauf (Eds.), *Language planning and policy language planning in local contexts* (pp. 3–17). Clevedon: Multilingual Matters.

Lubbe, A. (2020). *Cooperative learning-embedded assessment: Implications for students' assessment literacy and self-directedness in learning.* Unpublished doctoral thesis, North-West University, Potchefstroom.

Makalela, L. (2013). Translanguaging in kasi-taal: Rethinking old language boundaries for new language planning. *Stellenbosch Papers in Linguistics Plus, 42,* 111–125.

Makalela, L. (2018). Introduction: Shifting lenses. In L. Makalela (Ed.), *Shifting lenses: Multilanguaging, decolonisation and education in the global South* (pp. 1–8). Cape Town: Centre for Advanced Studies of African Society (CASAS).

Mathole, Y. (2016). Using content and language integrated learning (CLIL) to address multilingualism in South African schools. *European Journal of Language Policy, 8*(1), 57–77.

Mentz, E., Bailey, R., Havenga, M., Breed, B., Govender, D., Govender, I., et al. (2012). The diverse educational needs and challenges of information technology teachers in two Black rural schools. *Perspectives in Education, 30*(1), 70–78.

Merriam, S. B. (2009). *Qualitative research: A guide to design and implementation.* San Francisco, CA: Jossey-Bass.

Morrow, W. (2007). *Learning to teach in South Africa.* Cape Town: HSRC Press.

Ndebele, H. (2014). Promoting indigenous African languages through information and communication technology localisation: A language management approach. *Alternation Special Issue, 13,* 102–127.

Njobe, M. P. (2007). *Understanding the influence of a second language on the academic performance of learners in Information Technology: A case study of isiZulu-speaking english second language learners in KwaZulu Natal.* Unpublished master's dissertation, Durban University of Technology, Durban.

Olivier, J. A. K. (2011a). *Accommodating and promoting multilingualism through blended learning.* (Unpublished PhD thesis). North-West University, Vanderbijlpark.

Olivier, J. (2011b). Accommodating multilingualism in IT classrooms in the Free State province. *Southern African linguistics and applied language studies, 29*(2), 209–220.

Olivier, J. (2013). The accommodation of multilingualism through blended learning in a high school environment. *Perspectives in education, 31*(4), 43–57.

Olshtain, E., & Nissim-Amitai, F. (2004). Curriculum decision-making in a multilingual context. *International Journal of Multilingualism, 1*(1), 53–64.

Orman, J. (2008). *Language policy and nation-building in post-apartheid South Africa.* Dordrecht: Springer.

Saldaña, J. (2011). *Fundamentals of qualitative research*. Oxford: Oxford University Press.

Seemiller, C., & Grace, M. (2019). *Generation Z: A century in the making*. Abingdon: Routledge.

Skutnabb-Kangas, T. (1988). Multilingualism and the education of minority children. In T. Skutnabb-Kangas & J. Cummins (Eds.), *Minority education* (pp. 9–44). Clevedon: Multilingual Matters.

Van Deur, P. (2017). *Managing self-directed learning in primary school education: Emerging research and opportunities*. Hershey: IGI Global.

Venter, C. (2016). *The challenge of authentic Computer Applications Technology education in practice (Unpublished master's dissertation)*. Pretoria: University of Pretoria.

Verhoef, M. (1998). 'n Teoretiese aanloop tot taalgesindheidsbeplanning in Suid-Afrika. *South African Journal of Linguistics, 16*, 27–33.

Wright, S. (2016). *Language policy and language planning: From nationalism to globalisation*. Houndmills: Palgrave Macmillan.

Zhao, Shouhui, & Baldauf, R. B. (2012). Individual agency in language planning: Chinese script reform as a case study. *Language Problems and Language Planning, 36*(1), 1–24. https://doi.org/10.1075/lplp.36.1.01zha.

Jako Olivier is the holder of the UNESCO Chair in Multimodal Learning and Open Educational Resources and is a professor of Multimodal Learning in the Faculty of Education at North-West University, South Africa. His research, within the Research Unit Self-Directed Learning, focuses on self-directed multimodal learning, open educational resources, multiliteracies, blended and e-learning in language classrooms as well as multilingualism in education. He currently holds a Y rating from the National Research Foundation and was awarded the Education Association of South Africa's Emerging Researcher Medal in 2018. In addition to recently editing a book on self-directed multimodal learning, he has published numerous articles and book chapters at national and international levels.

Using Interactive Apps to Support Learning of Elementary Maths in Multilingual Contexts: Implications for Practice and Policy Development in a Digital Age

Nicola J. Pitchford, Anthea Gulliford, Laura A. Outhwaite, Lanaya J. Davitt, Evalisa Katabua, and Anthony A. Essien

Abstract Interactive apps are becoming increasingly popular in supporting the learning of elementary maths in primary schools internationally. In this chapter, we consider how educational apps might support learning in multilingual contexts. We describe three empirical studies conducted with bilingual children in primary schools in South Africa, the United Kingdom (UK), and Brazil: two draw on qualitative teacher data and one employs a group design to investigate the app's effects on children's learning. We report evidence that supports consideration of how multilingual presentation of this technology can aid learning in elementary maths. We found that the app can support children's choice of language of instruction and their learning in differing bilingual contexts. Implications for app-developers in improving access in multilingual contexts are considered. Attention is drawn to the pedagogical features of apps that facilitate multilingual learning. Finally, we consider the findings from these three countries within the wider context of multilingual instructional practices and policies for linguistic transfer. In so doing, we offer guidance for practitioners and policy-makers on app-based maths learning in multilingual educational

N. J. Pitchford (✉) · A. Gulliford · L. A. Outhwaite · L. J. Davitt
School of Psychology, University of Nottingham, Nottingham, UK
e-mail: nicola.pitchford@nottingham.ac.uk

A. Gulliford
e-mail: anthea.gulliford@nottingham.ac.uk

L. A. Outhwaite
e-mail: l.outhwaite@ucl.ac.uk

L. J. Davitt
e-mail: lanaya.davitt@nottingham.ac.uk

L. A. Outhwaite
Institute of Education, University College London, London, UK

E. Katabua · A. A. Essien
School of Education, University of the Witwatersrand, Johannesburg, South Africa
e-mail: anthony.essien@wits.ac.za

© The Author(s), under exclusive license to Springer Nature Switzerland AG 2021
A. A. Essien and A. Msimanga (eds.), *Multilingual Education Yearbook 2021*, Multilingual Education Yearbook,
https://doi.org/10.1007/978-3-030-72009-4_8

settings, considering aspects of classroom implementation and the optimisation of learning through understanding student languages.

Keywords Interactive apps · Bilingual/multilingual learners · Language proficiency · Translanguaging

1   Introduction

Educational apps are increasingly popular in supporting the learning of elementary maths by young children, especially the early years of primary school (Drinkwater, 2013; Pilli & Aksu, 2013; Zhang et al., 2015; Roberts & Spencer-Smith, 2019). When available in multiple languages, educational apps can offer a unique opportunity for learning maths in the child's home or preferred language(s), and as such, can be used in multilingual contexts. High-quality apps can increase access to education and boost achievement (Department for Education [DfE], 2019), however, technology alone will not equal success. To guide policy-makers, we need an evidence-base that considers barriers and enabling factors for successful implementation which impact learning outcomes (Outhwaite et al., 2019). For app-based maths learning, influential factors may include the child's cognitive development, socio-economic status, and linguistic skills especially in the language of instruction (Strand & Hessel, 2018), and teacher and classroom-level implementation practices, including beliefs and values of school leaders about the use of educational apps in their classrooms (Outhwaite et al., 2019).

2   Our Research

This chapter focuses on understanding how to optimise learning through implementation of app-based maths instruction in multilingual classrooms. Since 2013, we have been evaluating app-based maths learning environments in different countries, with monolingual and bilingual children. Our research focuses on a child-centred maths app, with proven efficacy developed by the UK-based, not-for-profit organisation, *onebillion*. This app provides the same curriculum content in multiple languages so offers a unique, cost-effective opportunity to support children in multilingual contexts to learn elementary maths in their home language, and/or language of instruction. We report the first cross-cultural comparison of using this app to supplement instruction in elementary maths across three countries—South Africa, the UK, and Brazil—each with different language contexts in the classroom and different policies for multilingual instruction. Aspects of the three studies are outlined in this chapter. It must be noted that the studies were adapted in each of the research contexts and as such, even though there were commonalities between contexts, we do not assume our work to be a replication study across the three contexts. However, the overarching question for the three studies was: How can learning of elementary maths be optimised when implementing the *onebillion* maths app in multilingual contexts? We conclude this chapter with implications for app-developers, practitioners, and policy-makers.

3 App Features and Content

The maths app used in our research offers a personalised numeracy software designed to deliver child-centred tuition through interactive picture, audio, and animation formats with clear objectives, instructions, and immediate formative feedback, provided by an on-screen teacher, consistent for all users (see Fig. 1). The software allows each child to have access to their own personal profile, labelled with their name, in which their progress is recorded.

Children work through the app individually with headphones, at their own pace, and can repeat instructions and activities as often as they desire. The app covers a range of topics in elementary maths, including counting, understanding the relations between numbers, basic numerical operations, and pattern and shape recognition. Within each topic there are a series of activities that introduce the child to a specific maths concept through small, sequenced units of information taught explicitly which provides the child with targeted practice to support skill acquisition. To complete a maths topic within the app, children need to achieve 100% pass rate on a 10-item quiz at the end of each topic. The quizzes are designed to assess children's knowledge of the mathematical concepts covered in the topic activities. Children are awarded with a star and certificate within the app when they complete all of the activities within a topic and answer all the quiz questions correctly.

These app features align with the principles of active, engaged, meaningful, and socially interactive learning (Hirsh-Pasek et al., 2015), the latter offered through the in-app teacher providing feedback and through pedagogical scaffolding from the adults present. The app content is age-appropriate and visually appealing. The app draws on features of direct instruction and retrieval-based learning, with step-wise progression through tasks, contingent feedback, and rehearsal to 'mastery'

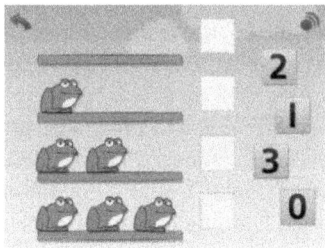
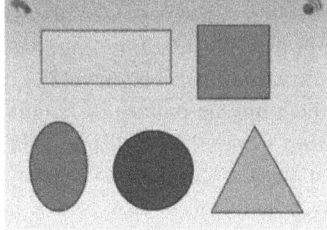

'Put the numbers in place'   'Choose the triangle'

Fig. 1 Greyscale images of example activities included in the *onebillion* maths app (actual display is coloured) (Courtesy of *onebillion*)

level (Gulliford & Miller, 2015). The apps also provide a self-paced and therefore individualised learning environment. These features are known to support learning for all children (Grimaldi & Karpicke, 2014; Kirschner et al., 2006; Slavin & Lake, 2008). The app also enables the same maths instruction and content to be presented in 52 different languages, so has potential to be a useful resource in multilingual classrooms.

4   Apps in a Bilingual Learning Context

The implementation of the *onebillion* app in bilingual and multilingual contexts draws upon the principle of cross-linguistic transfer (Cummins, 2017), offering explicit support for either the home language, language of instruction, or both. This personalised approach is compatible with a heteroglossic learning environment (Gandara & Randall, 2019). As a structured resource, the app is unable to offer flexible and dialogic pedagogical practices associated with a translanguaging perspective (García & Lin, 2017). Nevertheless, it can support children's retrieval of linguistic knowledge relevant to curriculum tasks presented in other languages, and could therefore be considered to be compatible with García's (2009) notion of translanguaging, whereby the child's receptive language may be different from their expressive language, and their language resources are activated through pedagogical features of the educational environment (Mizzi, 2019). While the app presents content in one 'language' at a time, the child is able, through flexible implementation, to activate their various language repertoires.

Technology can be used to scaffold vocabulary instruction (Daniel & Cowan, 2012) and *contextual* proficiency by supporting children to employ their existing language skills in the application of vocabulary in science contexts (Oyoo, 2017). The interactive design features of the *onebillion* app provide a good degree of contextual support, or scaffolding, optimising the acquisition of linguistic knowledge needed to navigate the cognitive demands of each task (Wright & Baker, 2017). Visual task demonstrations and verbal instructions are provided by the on-screen teacher, combined with interactive virtual objects, verbal labels, and numerical representations. Clear learning objectives and simple step-wise task instructions that children can repeat may also reduce the cognitive demands of the activity (Kirschner et al., 2006), aiding linguistic engagement when the app is delivered in the child's non-dominant language(s).

As a child's unique pattern of language experiences and pragmatic usage influences how they respond in multilingual instructional contexts (Planas & Setati, 2009), the app can support children to identify their optimal learning conditions, and to choose which language(s) they prefer, in which learning context (Grosjean, 2010). The app can therefore potentially promote *autonomy*, and thereby self-determination of the child (Deci & Ryan, 2012). As such, the app can support the child's linguistic identity, functioning as an aid to the child's sociocultural identity as a dual or multiple language learner (Hornberger & Link, 2012; Setati et al., 2008). This is especially important for children whose home languages are in the minority among their educational community, with few peers or adults able to aid the linguistic exchanges.

5 App Curriculum and Implementation

The content of the *onebillion* app is based on the elementary maths curriculum of England, so consideration is needed to the extent to which the app content aligns with the elementary maths curriculum of the country in which it is being implemented and the accuracy by which language within the app, especially mathematical terms, has been translated. In particular, the mathematics register needs to be appropriate for the language of instruction to optimise teaching and learning for mathematical concepts. We explore these issues in the study based in South Africa by examining linguistic features of the app content in relation to the expectations of teachers and children, both users of multiple languages and dialects.

As this app is available in multiple languages it can be implemented in the classroom in either the home language (if available in the software), language of instruction (if different to the home language), or both—by interleaving languages when teaching new concepts. These different language implementations enable exploration of how best to teach elementary maths with interactive apps in multilingual classrooms. Furthermore, for teachers to embrace using this app in their classrooms, it is important to observe their ability to implement the app effectively and identify the benefits and hurdles they experience when using the app as a teaching aid. In a study based in the UK, we compare these three distinct app implementation models in a low socio-economic multilingual context, to enable insight into how to optimise the app implementation for bilingual children and their teachers.

It is also necessary to evaluate how well bilingual and multilingual children learn elementary maths with this app when delivered in different languages, and the extent to which their proficiency in the language of instruction may influence how they progress through the app content. In a study conducted in Brazil we examine app implementation in a dual-immersion context (Gomez et al., 2005), specifically exploring the efficacy of teaching elementary maths with the *onebillion* app delivered in either Brazilian Portuguese or English in unbalanced bilingual children.

For each study, ethical approval was secured by the local commissioning body and parental consent was gained according to the local ethical requirements. Comparing these three studies, conducted in different countries with different language contexts, provides insight into how best to utilise interactive apps to support the learning of elementary maths in primary school classes comprised of children that speak more than one language.

6   South African Study

There are 11 official languages in South Africa exclusive of sign language (Stein, 2017). Given its multilingual nature coupled with apartheid history, the language in education policy (Department of Education, 1997) has made provision for children in the early years of education (ages 5–9) to receive instruction in their home language. This is believed to support conceptual growth, as well as ensuring continuity between the child's home language and language of instruction, while fostering equal access to education for all (Adler, 2001; Phakeng & Essien, 2016). Current practice is for the language of instruction to change to English in fourth grade, at 10 years of age. However, many schools use English as the language of instruction in all grades, even though the majority of children can understand, talk, and write in one of the 10 official South African languages. Furthermore, parents are often influenced by past experiences of how language was used as an oppressive tool during apartheid, and therefore believe that mastery of English will provide their children with better opportunities, and hence prefer their children to be taught in English (Stein, 2017; Barwell et al., 2016; Setati, 2005).

The practicalities of how this language policy is orchestrated in schools raise several concerns. First, with the migration from rural to metropolitan cities and vice versa, it is likely that children in the early years of primary school will experience a change in the language of instruction from their home language because schools tend to use the most frequently spoken language in a given catchment area. Children who move geographical locations will have to switch the language of instruction, regardless of their proficiency in that language. Second, teachers are often not sufficiently equipped to deal with multilingual classes as, more often than not, their pre-service training does not pay attention to the complexity of teaching and learning in multilingual contexts (Essien, 2010). Third, even for experienced teachers, working in multilingual classrooms poses several dilemmas, as they must continuously support code-switching and mediation as well as transparency (Adler, 2001). This makes delivering high-quality instruction in the early years of primary school especially challenging.

Given these challenges, we explored the scope of the *onebillion* maths app in supporting primary school teachers in South Africa by providing instruction of elementary maths in the child's home language, which was isiZulu. We were particularly interested in exploring the experiences of teachers and children using the isiZulu version of the *onebillion* maths app, to see if the isiZulu language translation used in the app was adequate to support learning of mathematical concepts, and the extent to which app content provided sufficient coverage of the South African primary school mathematics curriculum.

The study was based in a rural state primary school in the Kwa-Zulu Natal Province of South Africa, where isiZulu is the predominant language. The isiZulu version of the *onebillion* maths app was introduced to first-grade children (age 6 years) to supplement traditional classroom instruction in mathematics. Several methodologies were employed to investigate the experiences of primary school teachers and children when using the isiZulu version of the *onebillion* maths app. First, semi-structured interviews were conducted with five teachers and seven first-grade children. Open-ended questions (see Excerpt 1) allowed participants to voice their experiences without subjection of the researcher's preconceived ideas (Creswell, 2012). Second, two observations of teaching and learning were conducted, one with traditional classroom instruction and another with instruction through the app. Each classroom observation comprised of around 20 children. The researcher assumed a non-participant observer role and recorded observations in written format. Third, curriculum documents were analysed to establish the quality of text integration within the *onebillion* app (Creswell, 2012). Hardman's (2008) framework was adapted to analyse the data generated into different themes and subcategories.

Excerpt 1 Examples of open-ended questions posed to teachers who used the isiZulu version of the *onebillion* maths app to support instruction of mathematics with first-grade learners.

Let's talk about the learners in your class. Do you feel that the app has helped you and the learners attain the mathematics better and if so how so?

Are there any isiZulu language issues in the app that you would like to bring to the fore?

Has the mathematics register improved? And have you seen the difference, how?

Three key findings emerged. First, teachers perceived the self-paced feature of the app—where children progress through app content at their own pace, enabling some children to outpace the sequence of instruction given in the traditional classroom setting—as advantageous both for children and teachers. Teachers expressed that exposure to mathematical terms taught in isiZulu by the app, prior to that content being taught in class, facilitated children's understanding and accelerated their learning during traditional classroom instruction as the teacher did not need to teach the isiZulu mathematical register before teaching the content. Second, teachers wanted to use the app to become fluent in English mathematical terms, which they perceived as simpler linguistically than the isiZulu equivalent. For example, the English word '*eight*' when translated into isiZulu is '*(ku)isishiyagalombili*'. Seemingly, some teachers find introducing linguistically complex isiZulu mathematics register to children daunting and would rather use simpler English terminology to teach new mathematical concepts. This may arise because teachers, especially pre-service teachers, are often not equipped with the skills to teach in the ever increasing multicultural, multilingual, and super-diversified classroom that typifies state primary schools in South Africa (Barwell, 2016; Essien, 2010). Third, some inconsistencies were identified between the isiZulu used in the app and the curriculum documents. These included (i) omission of the term zero in the *onebillion* app compared to the curriculum documents in which the terms '*ziro/uziro/okungekho*' were used and (ii) different terminology used by the app and curriculum documents to refer to the shapes rectangle and oval, as the app used the terms '*raktango*' and '*ovali*' whereas the curriculum documents used the terms '*unxande*' and 'ukusaqanda', respectively. Despite these differences, children were able to progress through the app and were able to follow instructions within the traditional classroom setting.

7 UK Study

Currently, 21% of the total school population in the UK is identified as bilingual (DfE, 2018), however, numbers of children who speak English as an Additional Language (EAL) vary widely between regions, local authorities, and schools. Within the EAL student demographic, over 50 different languages are spoken (with at least 200 students per language group), the most frequent being Urdu, Punjabi, Bengali, and Polish.

While bilingualism has positive associations with various outcomes (Adesope et al., 2010), and being a bilingual learner per se should not contribute to educational risk, children who speak EAL may nevertheless be at risk of lower educational achievement than their monolingual peers in the UK (Strand et al., 2015; Strand & Hessel, 2018). For EAL children in the Early Years Foundation Stage (ages 4–5 years), 64% reach 'at least the expected standard' whilst 72% of non-EAL children reach this marker (DfE, 2018). Risks to EAL learning include recent arrival to the UK (Hutchinson, 2018), level of English language proficiency (Demie, 2018), and high mobility, entitlement to free school meals, and living in low socio-economic status (SES) areas (Strand et al., 2015). Thus, within the UK context, it is important to enhance the educational trajectory for these young EAL children.

In the UK study, we sought to identify ways in which implementation of the *onebillion* maths app could be optimised for an early years multilingual context. We explored three different implementation models of the app to establish which was most feasible for teachers of early years children.

- Model 1 was implemented with children whose home language was English. As these were non-EAL children, they received the *onebillion* maths app in English only.
- Model 2 was implemented with EAL children whose home language was not available within the *onebillion* maths app. These EAL children received instruction with the maths app delivered in English only.
- Model 3 was implemented with EAL children whose home language was available in the *onebillion* maths app. Within this model, each topic was given to the child in their home language and was then repeated in English, hence app content was interleaved between the child's home language and language of instruction.

The study took place in a mixed gender, inner-city, mainstream primary school in England. The school population of 417 children included a high number of children who came from a low socio-economic background, as measured by entitlement to free school meals: 37% compared to a national average of 23%. By the end of Year 6 (ages 10–11 years), 71% of children in this school typically reach the expected standard, which is higher than the national average of 66%. The school has higher than average levels of EAL learners, with 46% of children having EAL status compared to a national average of 21%. Within the early years population at the school, 21 different home languages were spoken by EAL children, comprising European, Asian, and African languages.

Three teachers participated in this study who taught children in Early Years Foundation Stage and Year 1 classes (ages 4–6 years). Each of the teachers identified 10 children from their class who they considered to be most in need of additional support with learning maths to receive intervention with the *onebillion* maths app. The total sample of 30 children (16 females and 14 males, mean age = 5.6 years, age-range 4.6–6.3 years) consisted of 18 non-EAL children and 12 EAL children. The language composition of the EAL group included Polish (3), Arabic (2), Kurdish (2), Urdu (1), Czech (1), Telugu (1), Igo (1), and Romanian (1). Proficiency in English (Strand & Hessel, 2018) was measured for each of the EAL children using a parent rating scale, consisting of four questions taken from the language subscales of the 5–15 Revised questionnaire (Kadesjö et al., 2017). Parents completed the questionnaire with the support of a specialised EAL teaching assistant to ensure they fully understood the questions. Using this measure the EAL group was rated as having moderate proficiency in English (mean parental rating = 5.90, SD = 2.07; possible highest proficiency = 4, possible lowest proficiency = 12). There was no significant difference in the English proficiency of the EAL children who received app implementation Model 2 compared to app implementation Model 3 [$t(10)$ = .93, p = .38].

Teachers delivered the maths app to participating children for 10–30 min a day, for 4 days a week, over a 12-week intervention period. Across the intervention period, delivery of the different models was observed once a week using an Intervention Implementation Fidelity Checklist developed from Outhwaite et al. (2019), supplemented by additional observations. The evaluative criteria covered several factors, including level of teacher support given to children, practical learning environment for the intervention, and level of child engagement with the app. Additional observation notes captured any further barriers or facilitating factors that teachers faced when implementing the app.

At the end of the intervention period, teachers took part in semi-structured interviews regarding the ease and suitability of using the *onebillion* maths app in their classroom, as well as any outcomes they may have seen. Interview data were analysed using thematic analysis (Braun & Clarke, 2013) and were triangulated with the class observations, to identify themes pertaining to the different implementation models for the app. Table 1 captures a synthesis of this small-scale investigation of benefits and hurdles with the different app implementation models.

Across the three implementation models studied, teachers found the *onebillion* maths app valuable in enabling rehearsal of skills learned in the classroom. The flexible nature of self-paced app-based learning was valued, as in the South African study, and was seen to support the individualised trajectory for children with diverse sociolinguistic profiles. As highlighted in Table 1, the interleaved implementation model (Model 3), enabling children to access the app content alternatively between their home language and English, was enjoyed by most EAL learners. A minority did not enjoy this method of instruction, however. The data indicated these were children

Table 1 Synthesis of the three teachers' perceptions (T1, T2, T3) of the different implementation models of the *onebillion* maths app with early years children in the UK

Implementation model	Perceived benefits	Perceived hurdles
Model 1 English Language Only for Non-EAL Learners *(English = 18)*	Children can access extra instruction with the maths curriculum, with minimum impact on teacher time and resources. T1, T2	Time spent interacting with the app potentially limits time available for general classroom routines and instruction. T2
Model 2 English Language Only for EAL Learners *(Kurdish = 2; Telugu = 1; Igbo = 1)*	The maths app allows children to repeat instructions when learning new words or concepts. T1	Challenges in controlling content exposure as children are able to skip important topics. T1, T2, T3
Model 3 Interleaving Home Language and English for EAL Learners *(Polish = 3; Arabic = 2; Urdu = 1; Czech = 1; Romanian = 1)*	Most EAL children enjoy and benefit from learning maths in their home language. T2, T3	Technical issues relating to ease of switching between different language versions of the maths app. T1, T2, T3 Two children did not enjoy learning maths in their home language and required additional pedagogical support to understand the maths terminology of their home language. T2, T3

whose home language skills did not allow them sufficient access to the curriculum content within the app. These children needed additional pedagogical support in their home language to enable them to access the app content. This highlights the need for careful scaffolding to enable children to transfer learning in home languages to instruction in other languages (Oyoo, 2017).

8 Brazilian Study

Brazil has a rapidly growing economy (World Bank, 2018) but there is a large disparity in the provision of quality education between public and private sectors (Akkari, 2013). The emergence of bilingual schools, particularly in the private sector in Brazil in the 1980s, was historically, socially, politically, and economically weighted, and aimed to promote a globalised education through a bilingual curriculum. However, the level of language immersion varies greatly across individual schools (Fortes, 2017). Previous research examining the impacts of bilingual immersion programmes have typically focused on the North American experience, for example, English-French immersion in Canada and English-Spanish immersion in the United States of America (Baker et al., 2016). Very little research has focused on the Brazilian experience. Brazil also faces numerous educational challenges: 48%

of children do not attain a basic level of mathematics (UNESCO, 2011) and the influence of the language of instruction on mathematics is of particular concern for young children at the start of school (Haag et al., 2015).

As the *onebillion* maths app can deliver the same maths instruction in Brazilian Portuguese and English, this study examined if the app could provide an effective intervention to support maths development in bilingual classrooms. First, adequacy of the translation between the app content in English and Brazilian Portuguese was assessed by two reviewers: a native English speaker and a native speaker of Brazilian Portuguese and a second-language speaker of English. Brazilian Portuguese and English transcripts were randomly selected for maths activities within the app. All selected app transcripts were deemed to accurately convey the same meaning in both languages (except for one, where the meaning of the instruction given was changed in translation from English to Brazilian Portuguese).

To explore how best to implement the maths app within a bilingual context, we conducted a study in a Brazilian Portuguese-English bilingual immersion private school in Recife in the North-East of Brazil (Outhwaite et al., 2020). The participating school implemented a one-way dual language programme with 50/50 immersion (Gomez et al., 2005). Time allocated for instruction in each language was split equally across the school day. In the first half of the school day all subjects, including maths, were taught in English. The second half of the school day was dedicated to language skills, including reading and writing, which were taught in Brazilian Portuguese.

A total sample of 62 children aged 5–6 years in the first year of elementary school took part. As we were interested in how the language of instruction influenced children's learning of elementary maths with or without the interactive app, children's proficiency in Brazilian Portuguese and English was assessed with a teacher-rated, 7-item questionnaire given at pre-test. The questionnaire included four items on speaking, and one item on each of reading, writing, and listening, and was repeated for each child for Brazilian Portuguese and English. It was developed specifically for this study and was adapted from the Alberta Language and Development Questionnaire (ALDQ; Paradis et al., 2010). The ALDQ is a non-language and non-culturally specific parental questionnaire and assesses children's competencies in their first and second language. Teacher ratings showed that, as a group, participating children were unbalanced bilinguals as their proficiency in Brazilian Portuguese was stronger than their proficiency in English.

The study followed a non-randomised, quasi-experimental design with three groups: 23 children in Class 1 received the maths app in Brazilian Portuguese, 20 children in Class 2 received the maths app in English, and 19 children in Class 3 received their regular maths teaching instruction, which in this school was delivered in English. Children allocated to receive instruction with the *onebillion* maths app (Class 1 and Class 2) used the app for 20-minutes, four times a week, for 10 consecutive weeks. The app was implemented by the class teacher and teaching assistant and was given instead of a small group embedding mathematics activities used in standard practice.

To evaluate learning with and without the maths app, all children completed the Early Grade Mathematics Assessment—EGMA (RTI International, 2009), before and after the 10-week intervention period. Assessments were delivered on a one-to-one basis with each child in a quiet area, free from distraction, in the child's familiar school environment. All assessment instructions were delivered using an audio recording of a standardised script in Brazilian Portuguese. For items that required a non-verbal response, children responded by pointing (e.g. pointing to which number was bigger for two presented numerical digits). For tasks that required a verbal response, children could choose to respond in either Brazilian Portuguese or English. Children's progress through the maths app was also assessed by the number of topics completed, as this has been shown to correlate significantly with EGMA performance in a group of 116 grade 1 children (6–7 years) attending mainstream school in Malawi demonstrating that the more topics children passed the more their maths ability improved (Pitchford et al., 2018, p. 8).

Learning gains across the intervention period were calculated for each of the three instructional groups. Statistical analysis showed that in this bilingual immersion context, when children received instruction with the *onebillion* app, given in either Brazilian Portuguese or English, they made significantly more progress in learning elementary maths than children who received regular classroom instruction [one-way analysis of variance: F (2, 58) = 3.78, p = .029]. On average, over the course of the 10-week intervention period, children who received app-based maths instruction in Brazilian Portuguese gained 17% on EGMA, children who received app-based maths instruction in English gained 18%, whereas children who received regular mathematics instruction in English given by their class teacher gained 11%.

Interestingly, there was no significant difference in learning gains achieved between classes of children that received the maths app as measured by EGMA, although children who used the app in Brazilian Portuguese made more progress through the app, passing 14 topics on average, compared to children who used the app in English, who passed on average 12 topics. Moreover, proficiency in the language of instruction was shown to be positively associated with progress through the maths app, as indicated by the number of topics completed: children with greater proficiency in the language of instruction made significantly more progress through the app than those with less proficient language skills [$r = 0.36$, $p = .022$] but this relationship was not significant for learning gains, as measured by EGMA [$r = 0.25$, $p = .113$].

These results demonstrate that the *onebillion* app can be an effective tool for learning elementary maths for bilingual children, when delivered in either the child's first or second language. In addition, proficiency in the language of instruction was significantly associated with progress through the app (topics passed), but not with learning gains as measured by EGMA. This most likely reflects differences in the sensitivity of the two outcome measures as a positive correlation was found with both measures, demonstrating again the importance of language proficiency in accessing learning materials within the app (Strand & Hessel, 2018; Cummins, 2008). Compared to standard maths instruction delivered by class teachers, app-based maths learning might be constrained by the inability of the app to provide additional supportive and contextual cues, such as gestures, intonation, and concrete aids, which facilitate learning for bilingual children.

9   Implications for Future Research

These studies provide evidence that the *onebillion* app can make an additive and flexible contribution to bilingual learning environments of various configurations. It can be considered a useful tool to support teaching and learning in multilingual contexts, being one artefact in a bilingual education mosaic (de Wal Pastoor, 2005). Learners valued using the app in their home language(s), and selecting which language to use for this curriculum domain, drawing upon the complementarity principle (Grosjean, 2010) of different languages serving differing contexts for the individual (Planas & Setati, 2009).

One interpretation for the cross-linguistic transfer that the app supports, is that it offers facilitation with code-switching (Moschkovich, 2007). Alternatively, the translanguaging perspective, sees the child moving flexibly between their various language resources (Gandara & Randall, 2019). Evidence from the South African study, that children were seemingly not impeded by mismatches in the mathematical terminologies used in the app and classroom practice, provide some support for the translanguaging perspective. To enhance understanding of how linguistic activation and transfer may be occurring, further investigations are needed of how the learner is engaging with the app, in which languages, and with what contingencies, in relation to maths outcomes. Further finer-grained analysis of interactions within the app could support insights into how bilingual or multilingual children engage with the app content, and the mechanisms that support their linguistic and cognitive development within this instructional context.

10   Implications for App-Developers

Our studies suggest that app-developers should consider the role of language when designing app features and content. For example, while the *onebillion* app provided a level of contextual support, through matching visual and verbal information, interactive virtual objects, and numerical representations, some content may have required vocabulary beyond the child's current knowledge. These vocabulary challenges may have limited the learning progress of some children. To improve access to app-based learning for all children, developers could incorporate further context-embedded communication cues such as gestures, intonation, and concrete cues. Also, when translating apps between languages, care needs to be taken to ensure the translation is accurate, to maintain the meaning of terminology between languages. The mathematics register also needs to align with the curriculum of a given country, both in terms of accuracy and content. Finally, to facilitate a translanguaging learning environment, app-developers could enhance ease of interleaving languages of instruction within tasks, so languages are taught simultaneously by the app, and allow the learner a choice of preferred language, without requiring teachers to manually switch delivery of different languages within the app.

11 Implications for Practitioners

Our findings suggest that children benefit from sufficient proficiency in the language of instruction used within apps to access curriculum content and respond to instruction (Cummins, 2008). Teachers and parents wishing to use interactive apps with bilingual or multilingual children should therefore consider the child's proficiencies in the various languages spoken. When available in both the home language(s) and language of instruction, our research suggests it might be optimal to give bilingual or multilingual children experience of app-based maths learning in both languages and then allow children to decide which language they would prefer to use. Allowing bilingual or multilingual children to determine their preference might encourage positive engagement, as self-determination can increase intrinsic motivation to perform academic tasks (Deci & Ryan, 2012).

The South African study also showed that children are able to use the app without a precise match between terminology used by the app and teachers delivering the curriculum. This suggests that app-based maths learning can be appropriate even in contexts where many languages exist with diverse dialects and academic nomenclature. Moreover, children seemed able to adapt to this diversity is important in helping them use their home language(s) in the classroom, again supporting the sociocultural positioning of their language identity (Setati et al., 2008; Hornberger & Link, 2012), and potentially a translanguaging approach (García, 2009). On the other hand, such divergences may also warrant attention from instructors, where different 'versions' of the mathematics register become too distinct.

12 Implications for Education Policy

Our research shows that interactive maths apps available in multiple languages offer a flexible resource for education policy-makers. High-quality, multilingual, apps can provide a ready-made, step-wise, curriculum resource for implementation in the home language(s), language of instruction, or both. Policy-makers should bear in mind how to optimise the implementation of interactive apps for their particular setting, its languages, and community resources, for example, by considering parental engagement and availability of pedagogical scaffolding. In addition, policy-makers should examine existing research evidence supporting the effectiveness of interactive apps, and how they align to bilingual and multilingual theories, learning theory, and curricula. Such bespoke reflection is key to optimising any bilingual and multilingual education provision, however, some generic points for policy-makers emerge from this work. Our research suggests it is beneficial if:

- Children are able to access the maths curriculum in their home language(s).
- Children are able to interact with the maths curriculum in a way that allows them to rehearse skills at an individualised pace.
- Children are enabled to select their preferred language of instruction for learning.

- Teachers and parents develop a shared knowledge of the child's proficiencies in the languages understood and spoken, and in which contexts.
- Teachers and parents develop a shared knowledge of the child's cognitive skills in their home language(s), and provide appropriate scaffolding in the language of instruction to support the child's successful engagement with app content.

Finally, interactive apps may be viewed as part of the mosaic of multilingual educational policy: not the sole answer to bilingual and multilingual instruction, but forming a useful part of the response to challenges faced by educators. As seen in the South African context, implementation of multilingual practices under policy directives may present significant challenges for teachers, despite the language diversity in a country. Interactive apps can offer support for the fine-grained task of transferring learning from one language to another, enhancing the learning of a child in elementary maths.

References

Adesope, O. O., Lavin, T., Thompson, T., & Ungerleider, C. (2010). A systematic review and meta-analysis of the cognitive correlates of bilingualism. *Review of Educational Research, 80*(2), 207–245.

Adler, J. (2001). *Teaching and learning mathematics in multilingual classrooms.* Dordrecht, Netherlands: Kluwer Academic Publishers.

Akkari, A. (2013). Blurring the boundaries of public and private education in Brazil. *Journal of International Education and Leadership, 3*(1), 1–13.

Baker, D. L., Basaraba, D. L., & Polanco, P. (2016). Connecting the present to the past: Furthering the research on bilingual education and bilingualism. *Review of Research in Education, 40*(1), 821–883.

Barwell, R. (2016). Mathematics education, language and superdiversity. In A. Halai & P. Clarkson (Eds.), *Teaching and learning mathematics in multilingual classrooms: Issues for policy, practice and teacher education* (pp. 25–39). Rotterdam: Sense Publishers.

Barwell, R., Chapsam, L., Nkambule, T., & Setati-Phakeng, M. (2016). Tensions in teaching mathematics in contexts of language diversity. In R. Barwell, P. Clarkson, A. Halai, M. Kazima, J. Moschkovich, N. Planas, M. Phakeng, P. Valero, & M. Villavicencio Ubillús (Eds.), *Mathematics education and language diversity* (pp. 175–192). Switzerland: Springer.

Braun, V., & Clarke, V. (2013). Using thematic analysis in psychology. *Qualitative Research in Psychology, 3*(2), 77–101.

Creswell, J. (2012). *Educational research: Planning, conducting and evaluating qualitative and quantitative research* (4th ed.). Boston: Pearson Education.

Cummins, J. (2008). BICS and CALP: Empirical and theoretical status of the distinction. In B. Street & N. H. Hornberger (Eds.), *Encyclopedia of language and education.* New York, USA: Springer.

Cummins, J. (2017). Teaching for transfer in multilingual school contexts. In O. García, A. Lin, & S. May (Eds.), *Bilingual and multilingual education.* Cham, Switzerland: Springer.

Daniel, M. C., & Cowan, J. E. (2012). Exploring teachers' use of technology in classrooms of bilingual students. *GIST Education and Learning Research Journal, 6,* 97–110.

de Wal Pastoor, L. (2005). Discourse and learning in a Norwegian multiethnic classroom: Developing shared understanding through classroom discourse. *European Journal of Psychology of Education, 20*(1), 13.

Deci, E. L., & Ryan, R. M. (2012). Motivation, personality, and development within embedded social contexts: An overview of self-determination theory. In *The Oxford handbook of human motivation*. New York and Oxford: Oxford University Press.

Demie, F. (2018). English language proficiency and attainment of EAL (English as second language) pupils in England. *Journal of Multilingual and Multicultural Development, 39*(7), 641–653.

Department of Education. (1997). *Language-in-education policy*. Pretoria: Department of Education.

Department for Education. (2018). *Schools pupils and their characteristics 2018 local authority tables*. London: Department for Education. https://www.gov.uk/government/statistics/schools-pupils-and-their-characteristics-january-2018.

Department for Education. (2019). *Realising the potential of technology in education: A strategy for education providers and the technology industry*. https://assets.publishing.service.gov.uk/government/uploads/system/uploads/attachment_data/file/791931/DfE-Education_Technology_Strategy.pdf.

Drinkwater, D. (2013). *Brazilian government spends big on nearly 500,000 tablets for local teachers*. http://tabtimes.com/news/education/2013/11/27/brazilian-government-spends-big-nearly-500000-tabletslocal-teachers.

Essien, A. (2010). Mathematics teacher educators' account of preparing pre-service teachers for teaching Mathematics in multilingual classroom: The case of South Africa. *International Journal of Interdisciplinary Social Sciences [serial online], 5*(2), 33–44.

Fortes, L. (2017). The emergence of bilingual education discourse in Brazil: Bilingualisms, language policies, and globalizing circumstances. *International Journal of Bilingual Education and Bilingualism, 20*(5), 574–583.

Gandara, F., & Randall, J. (2019). Assessing mathematics proficiency of multilingual students: The case for translanguaging in the democratic Republic of the Congo. *Comparative Education Review, 63*(1), 58–78.

García, O. (2009). *Bilingual education in the 21st century: A global perspective*. Malden/Oxford: Blackwell/Wiley.

García, O., & Lin, A. (2017). Translanguaging in bilingual education. In O. García, A. Lin, & S. May (Eds.), *Bilingual and multilingual education* (pp. 117–130). Cham, Switzerland: Springer.

Gomez, L., Freeman, D., & Freeman, Y. (2005). Dual language education: A promising 50–50 model. *Bilingual Research Journal, 29*(1), 145–164.

Grimaldi, P. J., & Karpicke, J. D. (2014). Guided retrieval practice of educational materials using automated scoring. *Journal of Educational Psychology, 106*, 58–68.

Grosjean, F. (2010). *Bilingual: Life and reality*. Harvard University Press.

Gulliford, A., & Miller, A. (2015). Raising educational achievement: What can instructional psychology contribute? In T. Cline, A. Gulliford, & S. Birch (Eds.), *Educational psychology* (pp. 99–123). Routledge.

Haag, N., Roppelt, A., & Heppt, B. (2015). Effects of mathematics items' language demands for language minority students: Do they differ between grades? *Learning and Individual Differences, 42*, 70–76.

Hardman, J. (2008). Researching pedagogy: An activity theory approach. *Journal of Education, 45*, 65–95.

Hirsh-Pasek, K., Zosh, J. M., Golinkoff, R. M., Gray, J. H., Robb, M. B., & Kaufman, J. (2015). Putting education in "educational" apps: Lessons from the science of learning. *Psychological Science, 16*(1), 3–34.

Hornberger, N. H., & Link, H. (2012). Translanguaging and transnational literacies in multilingual classrooms: A biliteracy lens. *International Journal of Bilingual Education and Bilingualism, 15*(3), 261–278.

Hutchinson, J. (2018). *Educational outcomes of children with English as an additional language*. London: Education Policy Institute and the Bell Foundation.

Kadesjö, B., Janols, L.-O., Korkman, M., Mickelsson, K., Strand, G., Trillingsgaard, A., Lambek, R., Øgrim, G., Bredesen, A. M., & Gillberg, C. (2017). *Five-To-Fifteen-Revised* (5–15R). https://www.5-15.org/pdf/515_en-GB.pdf.

Kirschner, P. A., Sweller, J., & Clark, R. E. (2006). Why minimal guidance during instruction does not work: An analysis of the failure of constructivist, discovery, problem-based, experiential, and inquiry-based teaching. *Educational Psychologist, 41*(2), 75–86.

Mizzi, A. (2019, February). Researching translanguaging: Functions of first and second languages in Maltese mathematics classrooms. In *Eleventh Congress of the European Society for Research in Mathematics Education* (No. 19). Freudenthal Group; Freudenthal Institute; ERME.

Moschkovich, J. (2007). Using two languages when learning mathematics. *Educational Studies in Mathematics, 64*(2), 121–144.

Outhwaite, L. A., Gulliford, A., & Pitchford, N. J. (2019). A new methodological approach for evaluating the impact of educational intervention implementation on learning outcomes. *International Journal of Research & Method in Education, 43*(3), 1–18.

Outhwaite, L. A., Gulliford, A., & Pitchford, N. J. (2020). Language counts when learning mathematics with interactive apps. *British Journal of Educational Technology*. https://doi.org/10.1111/bjet.12912.

Oyoo, S. O. (2017). Learner outcomes in science in South Africa: Role of the nature of learner difficulties with the language for learning and teaching science. *Research in Science Education, 47*(4), 783–804.

Paradis, J., Emmerzael, K., & Duncan, T. S. (2010). Assessment of English language learners: Using parent report on first language development. *Journal of Communication Disorders, 43*(6), 474–497.

Phakeng, M., & Essien, A. A. (2016). Adler's contribution to research on mathematics education and language diversity. In M. Phakeng & A. A. Essien (Eds.), *Mathematics education in a context of inequity, poverty and language diversity* (pp. 1–6). Switzerland: Springer International Publishing.

Pilli, O., & Aksu, M. (2013). The effects of computer-assisted instruction on the achievement, attitudes and retention of fourth grade mathematics students in North Cyprus. *Computers & Education, 62*, 62–71.

Pitchford, N. J., Kamchedzera, E., Hubber, P. J., & Chigeda, A. (2018). Interactive apps promote learning in children with special educational needs. *Frontiers in Psychology, 9,* 262.

Planas, N., & Setati, M. (2009). Bilingual students using their languages in the learning of mathematics. *Mathematics Education Research Journal, 21*(3), 36–59.

Roberts, N., & Spencer-Smith, G. (2019). A modified analytical framework for describing m-learning (as applied to early grade Mathematics). *South African Journal of Childhood Education, 9*(1), a532. https://doi.org/10.4102/sajce.v9i1.532.

RTI International. (2009). *Early Grade Mathematics Assessment (EGMA): A conceptual framework based on mathematics skills development in children.* Prepared under the USAID Education Data for Decision Making (EdData II) project, Task Order No. EHC-E-02-04-00004-00.

Setati, M. (2005). Teaching mathematics in a primary multilingual classroom. *Journal for Research in Mathematics Education, 36,* 447–466.

Setati, M., Molefe, T., & Langa, M. (2008). Using language as a transparent resource in the teaching and learning of mathematics in a Grade 11 multilingual classroom. *Pythagoras, 67,* 14–25.

Slavin, R. E., & Lake, C. (2008). Effective programs in elementary mathematics: A best-evidence synthesis. *Review of Educational Research, 78*(3), 427–515.

Stein, N. (2017). *Chapter 11: Language in schools.* http://section27.org.za/; http://section27.org.za/wp-content/uploads/2017/02/Chapter-11.pdf.

Strand, S., & Hessel, A. (2018). *English as an additional language, proficiency in English and pupils' educational achievement: An analysis of local authority data.* Oxford/London/Cambridge: University of Oxford, Unbound Philanthropy, The Bell Foundation.

Strand, S., Malmberg, L., & Hall, J. (2015). *English as an additional language (EAL) and educational achievement in England: An analysis of the National Pupil Database.* Oxford: Department of Education, University of Oxford.

UNESCO. (2011). *State of Education in Brazil.* Paris, France: UNESCO.

World Bank. (2018). *An adjusting partnership: World Bank group support to Brazil 2011-2018.* https://www.worldbank.org/en/results/2018/10/01/world-bank-group-support-brazil-2011-2018.

Wright, W., & Baker, C. (2017). Key concepts in bilingual education. In O. García, A. M. Y. Lin, S. May, & N. Hornberger (Eds.), *Bilingual and multilingual education* (pp. 65–80). Springer.

Zhang, M., Trussell, R. P., Gallegos, B., & Asam, R. R. (2015). Using math apps for improving student learning: An exploratory study in an inclusive fourth grade classroom. *TechTrends: Linking Research and Practice to Improve Learning, 59*(2), 32–39. https://doi.org/10.1007/s11528-015-0837-y.

Nicola Pitchford is a Professor of Psychology at the University of Nottingham. Her research expertise lies in developmental psychology and education. Nicola conducts basic, clinical, and applied research to investigate scholastic progression over childhood. She works at the interface of theory and practice, collaborating with academics from different disciplines and working alongside practitioners and professionals from a diverse range of fields to ensure her research secures maximum benefits for key users and stakeholders. partnership with the *Voluntary Service Overseas*, Nicola is leading an international programme of research exploring the use of innovative digital technologies to support the acquisition of basic numeracy and literacy skills by primary school children in Malawi, the UK, and other countries worldwide. She is a member of several advisory boards for organisations, including the EdTech Hub, that are using digital technologies to enhance learning of core foundational skills by young children.

Anthea Gulliford works in the School of Psychology at the University of Nottingham. She has extensive experience of training applied psychologists, having been a Programme Director to doctoral educational psychology programmes for many years. She is a chartered and registered Educational Psychologist and also works in Nottingham City Council's Children's Services. She holds a background in early years and bilingual and multilingual education. Anthea's work focuses on inclusion and equity through supporting the needs of vulnerable learners, including those excluded from school, those at risk of low achievement, those with social and emotional mental health needs, and deaf and hearing-impaired children and young people. Her research and training specialisms include the application of interpersonal skills in school psychology consultation; behaviour change models; group dynamics and organisational change; and applied research methods for educational psychology.

Laura Outhwaite is a Research Fellow in the Centre for Education Policy and Equalising Opportunities at UCL Institute of Education. Her work combines basic and applied research in psychology and education to understand the design, impact, and scaling of educational interventions. Her research focuses on mathematics, early years, and educational technology in different educational and cultural contexts. Prior to joining UCL, Laura completed her Ph.D. at the University of Nottingham under the supervision of Prof Nicola Pitchford and Anthea Gulliford. Her Ph.D. thesis evaluated the use of interactive maths apps to support early mathematical development in UK and Brazilian primary school children with a mixed-methods approach. This work provided the initial evidence for a large-scale evaluation trial of the maths app intervention in UK primary schools funded by the Education Endowment Foundation. Laura has also previously worked as a Research Mentor for the EDUCATE project at the UCL Knowledge Lab.

Lanaya Davitt is a postdoctoral research student at the School of Psychology, University of Nottingham, working under the supervision of Professor Nicola Pitchford and Anthea Gulliford. Her research focuses on using digital technology to support early years learning for children who speak English as an Additional Language. Her work takes a mixed methods approach: qualitative research focuses on teacher perspectives of implementing maths apps in multicultural classrooms whilst quantitative research evaluates the effectiveness of learning when using interactive apps in multilingual classroom environments. She is also exploring factors that underlie maths ability for children who speak more than one language to understand how best to support children with EAL through personalised learning programmes.

Evalisa Katabua is a middle school Mathematics Educator in Johannesburg, South Africa. She holds a Master's degree in Education from the University of Witwatersrand. Her interest as a researcher includes the intersection of language/multilingualism and mathematics which formed the basis of her Master's Degree. Evalisa was the recipient of the award for the best Master's student in the field of Education in 2019. Before joining the Mathematics Education sector, Evalisa worked in the NGO sector where she was exposed to primary and high school education as well as other humanitarian interventions.

Anthony A. Essien is an Associate Professor and the Head of the Mathematics Education Division at the University of the Witwatersrand, South Africa. He is a series editor of the book series *Studies on Mathematics Education and Society*. His field of research is in mathematics teacher education in contexts of language diversity. He is also a current member of the International Committee (Board of Trustees) for the International Group for the Psychology of Mathematics Education (IGPME). Anthony also served as an associate editor of *Pythagoras*, the academic journal of the Association for Mathematics Education of South Africa, for 11 years. In addition to his background in mathematics education, Anthony also has a background in Philosophy.

Noticing Multilingual and Non-dominant Students' Strengths for Learning Mathematics and Science

Salvador Huitzilopochtli, Julianne Foxworthy Gonzalez,
Judit N. Moschkovich, Sam R. McHugh, and Maureen A. Callanan

Abstract This chapter examines the strengths multilingual and non-dominant learners bring to school for learning mathematics or science. We connect research on learners' strengths from two sets of literature: research in mathematics education and in science education. The examples illustrate how research in mathematics and science education can inform policy and teaching in multilingual classrooms. This research assumes that learners from multilingual communities bring strengths, not deficits, to classrooms. We provide and illustrate three recommendations for effective policies and teaching: (1) noticing learners' strengths, (2) recognizing mathematics and science practices, and (3) expanding what counts as practices in STEM disciplines. We use examples from previously published research in United States classrooms to illustrate how to notice the strengths multilingual learners bring to learning math and science, recognize practices associated with STEM disciplines in student contributions, and expand what we include in such practices. Although the examples are drawn from classroom-based research in the United States, those findings have important implications that extend to other settings (communities, nations, or learning environments).

Keywords Equity · Strengths · Science · Mathematics · STEM practices · Language diversity

S. Huitzilopochtli (✉) · J. Foxworthy Gonzalez · J. N. Moschkovich (✉) · S. R. McHugh · M. A. Callanan
University of California, Santa Cruz, CA, USA
e-mail: shuitzil@ucsc.edu

J. N. Moschkovich
e-mail: jmoschko@ucsc.edu

J. Foxworthy Gonzalez
e-mail: jfoxwort@ucsc.edu

S. R. McHugh
e-mail: srmchugh@ucsc.edu

M. A. Callanan
e-mail: callanan@ucsc.edu

A. A. Essien and A. Msimanga (eds.), *Multilingual Education
Yearbook 2021*, Multilingual Education Yearbook,
https://doi.org/10.1007/978-3-030-72009-4_9

1   Introduction

This chapter focuses on noticing the strengths of multilingual and non-dominant learners for science, technology, engineering, and mathematics (STEM). We use a Vygotskian approach to learning and teaching that emphasizes learners' potential, not the mistakes they make. Approaches focusing on misconceptions and errors have been shown to be insufficient to support student learning (Hammer, 1996; Moschkovich, 1998, 1999a; Smith et al., 1994). Thus, it is important that teaching and policy for multilingual and non-dominant STEM students include strengths. Most importantly, deficit models of multilingual learners often neglect to notice any strengths that these learners bring to STEM classrooms (Barwell et al., 2017). This paper illustrates the strengths that multilingual learners bring for learning mathematics or science in school. We connect research in mathematics education to research in science education (in the United States), exploring these two sets of complementary findings. The chapter uses five examples from previously published research in United States classrooms to illustrate three recommendations for policy and practice.

A focus on learners' strengths can inform policy and teaching by first assuming that learners bring strengths to STEM learning, not deficits (Aguirre et al., 2013; Moschkovich, 2002). Beyond that assumption, we make three recommendations for policy and practice to: (1) notice learners' strengths (Mason, 2002; Watson, 2009), (2) recognize practices associated with the disciplines of mathematics or science[1] (Rosebery et al., 1992; Warren et al., 2001) in students' contributions, and (3) expand what counts as STEM practices. Although noticing students' strengths is necessary, it is only a first step. Practitioners need to not only notice strengths but also notice the disciplinary practices in what students do or say. Only by recognizing disciplinary practices in students' contributions can practitioners build on strengths and support students as they develop further disciplinary expertise.

Language policies facilitate or constrain multilingual students' access to rigorous STEM coursework (NASEM, 2018). Classifying students by proficiency in the language of instruction can have unintended consequences. For example, if students are placed in STEM courses according to their language proficiency, this can lead to systematic exclusion through "tracking" (NASEM, 2018). Such placement practices are based on beliefs that language must be mastered before students can engage with content or that language is learned separately from content. These unproductive beliefs (Faltis & Valdés, 2016) undergird language policies and practices in STEM education. Instead, policies should assume that, given appropriate conditions, multilingual students learn at least as well as their monolingual peers (Barwell et al., 2017).

We use bilingual, multilingual, and non-dominant to refer to learners from communities whose members speak one (or more) language(s) (or language varieties) different from the language of instruction (LOI). We use non-dominant (Gutiérrez,

---

[1]We will call these STEM practices, disciplinary practices, science practices, or mathematical practices.

2008) to include learners from marginalized communities and acknowledge power issues for communities and learners which may or may not be labeled multilingual. We use *multilingual* to emphasize student competencies instead of deficiencies (i.e., English learners or learners of the LOI). Valdes-Fallis' (1978) definition of a bilingual speaker is "the product of a specific linguistic community that uses one of its languages for certain functions and the other for other functions or situations" (p. 4). Policy and instruction should leverage what students have, not focus on what they do not know (Faltis & Valdés, 2016). Equitable teaching practices for multilingual learners need to shift from focusing on perceived deficits to uncovering, honoring, and building on students' strengths, in particular the "repertoires of practices" (Gutiérrez & Rogoff, 2003) students bring to classrooms. This chapter provides examples of noticing such strengths, recognizing STEM practices in student contributions, and expanding what counts as STEM practices. Only then, after these three recommendations are met, can policy and teaching build on these students' strengths.

2 Theoretical Framing

The first recommendation, noticing student strengths, depends on noticing as a practice. We draw on *professional vision* (Goodwin, 1994) to frame noticing student strengths. Louie (2018) frames teacher noticing as a teaching practice laden with the values of the larger educational system. Noticing is not neutral but has culturally based affordances and constraints (Goodwin, 1994; Louie, 2018). Our theoretical framework connects noticing student strengths to STEM practices. Multilingual students' strengths can enrich learning opportunities only when these strengths are noticed, when STEM practices in student contributions are recognized, and when we expand what counts as STEM practices. Only then can teaching build on those strengths.

Teachers learn to notice (i.e., attend, interpret student contributions) in a variety of ways. For example, teachers can focus on students' emerging reasoning or on errors, misconceptions, and perceived deficits. Louie (2018) emphasizes that noticing is socially constructed, not politically neutral, so whether teachers privilege reasoning or misconceptions has consequences for learning. As teachers refine their noticing practices, they reproduce ways of noticing that privilege certain students over others (Louie, 2018). Teachers are inculcated into noticing practices and influenced by the educational culture. Often, noticing practices focus on what students do not know, using a deficit lens. Such deficit views negatively impact students' classroom experiences, course placements, and opportunities to learn STEM. This intellectual, symbolic, and epistemological violence can have material consequences on student outcomes (Martin, 2019; NASEM, 2018).

Teachers must notice students' strengths and also recognize STEM practices in students' contributions. Professional vision can expand so that student strengths are not rendered irrelevant by narrowly defined practices. Educators need to both notice strengths and recognize STEM practices to provide opportunities for students to engage in STEM disciplinary practices in ways that build on student strengths.

We start with the assumption that students bring strengths (Gholson et al., 2012; Martin, 2019) and linguistic competence (Martínez & Mejía, 2020) to learning STEM in school. We make three recommendations for connecting students' strengths to STEM practices through the professional vision of teachers (Louie, 2018). First, educators should notice students' strengths, rather than the errors of "imperfect language" (Faltis & Valdés, 2016; Moschkovich, 2013). Second, educators should recognize the STEM practices in students' contributions. Third, educators should expand what "counts" as mathematics and science practices—so that those strengths are valued as crucial knowledge, expanding the unnecessarily narrow views of STEM practices (Bang et al., 2012). This expansion "desettles" (Bang et al., 2012; Harris, 1995) expectations so that community knowledge is valued alongside traditional definitions of STEM disciplinary knowledge or practices.

Our examples of particular strengths for STEM learning and disciplinary practices are not a prescription for teaching students from any particular community or group. Cultural practices vary according to the historical context, goals, and purposes of a community (Gutiérrez & Rogoff, 2003). Therefore, one cannot assume that the cultural practices of one group will necessarily be the same in another group that shares the same heritage.

We use the United States mathematics standards, the Common Core State Standards for mathematics (CCSS, 2010), and United States science standards, the Next Generation Science Standards (NGSS, 2013) as current policy embodiments of the STEM practices that should be available to students in school. Although these standards are informed by research on STEM professional practices, the standards do not capture everything important about STEM learning, nor are these standards assessment tools for student learning. The standards serve only as a shorthand for the disciplinary practices that researchers, practitioners, and policy makers in the United States agreed are central foci for STEM teaching and learning.

We focus on mathematical practices emphasized in the CCSS (2010), such as constructing arguments, reasoning abstractly, generalizing from mathematical structure, and modeling. We also focus on science practices emphasized by the NGSS (2013), including asking questions, analyzing and interpreting data, and arguing with evidence. These practices overlap with STEM professional practices and are recommended for classroom STEM instruction.

3   Mathematics Examples

Although traditional approaches focused on mastering mathematical procedures, we start with a broader definition of mathematical proficiency (Kilpatrick et al., 2001) and add mathematical practices (Moschkovich, 2013; Schoenfeld, 1992). Adding mathematical practices provides students opportunities to engage in the activities that mathematicians or those who use mathematics actually use, e.g., describing mathematical objects (examples 1 and 2) or making inferences from data (example 3). The following examples illustrate our three recommendations. Examples 1 and

2 describe students' strengths and show how we can recognize STEM practices in students' contributions. We use the third example to illustrate what is meant by expanding what counts as STEM practices.

3.1 Math Example 1: Noticing Students' Strengths in Abstracting and Generalizing

This excerpt is from a Grade 3 (eight and nine years old) classroom with multilingual students in an urban elementary school in California (Moschkovich, 1999b). Students received instruction in both English and Spanish; this lesson was part of a geometry unit on classifying shapes. The teacher began by holding up a rectangle and asking students to describe it (Moschkovich, 1999b, p. 13).

Eric: A rectangle has…two…short sides, and two…long sides.
Teacher: Two short sides and two long sides. Can somebody tell me something else about this rectangle? If somebody didn't know what it looked like, what, what…how would you say it?
Julian: Paralel(o). [holding up a rectangle]
Teacher: It's parallel. Very interesting word. Parallel, wow! Pretty interesting word, isn't it? Parallel. Can you describe what that is?
Julian: Never get together. They never get together. [runs his finger over the top length of the rectangle]
Teacher: What never gets together?
Julian: The parallela…the…when they go, they go higher [runs two fingers parallel to each other first along the top and base of the rectangle and then continues along those lines] they never get together.
Antonio: Yeah!
Teacher: Very interesting. The rectangle then has sides that will never meet. Those sides will be parallel. Good work. Excellent work.

Several strengths are evident in these contributions. First, Julian used gestures and objects to support his claim, making these strengths for communicating mathematically. He also used his first language (pronouncing "paralelo" in Spanish) to support his participation in this mathematical discussion. Instead of translating that word to English, he used the Spanish word. Next, he used everyday language to describe a property of parallel lines. Even though his claim "they never get together" is not formal, it does communicate a correct mathematical idea. There were also two mathematical practices in Julian's contributions. Julian was abstracting, describing an abstract property of parallel lines (one cannot see where lines do not "get together." He was also generalizing, saying that parallel lines will never meet, not only today or tomorrow, or here in this classroom, but never.

In this classroom discussion, the teacher did not correct Julian's English or object to his use of the Spanish word "parallela," in contrast to policies that restrict classroom talk to the language of instruction or teaching practices that focus solely on the mastery of vocabulary. Instead, he *revoiced* (O'Connor & Michaels, 1993) Julian's comments, asked questions to discover what Julian meant, and focused on the mathematical content, the particular features of parallel lines. Listening to students' contributions is the essential first step in noticing. Revoicing can build on students' own use of mathematical practices, or a student contribution can be revoiced to reflect new mathematical practices (Moschkovich, 2015). In this case, the teacher's revoicing made Julian's claim more precise, introducing a new mathematical practice: attending to the precision of a claim. The teacher's claim, "The rectangle then has sides that will never meet. Those sides will be parallel," is more precise because it refers to the sides of a quadrilateral, rather than any two parallel lines.

3.2 Math Example 2: Recognizing Mathematical Practices When Comparing Lines

The transcript below is from an interview with two Grade 9 students (14 and 15 years old) conducted after school (Moschkovich, 2011). The students had been in mainstream, English-only mathematics classrooms for several years. One student, Marcela, had some previous mathematics instruction in Spanish. The students were working on the problem in Fig. 1 after they had worked on problems with positive slopes greater and less than 1.

The students had graphed the line $y = -0.6x$ by hand on paper (Fig. 2) and were

8a. If you change the equation y=x to y=-0.6x, how would the line change?

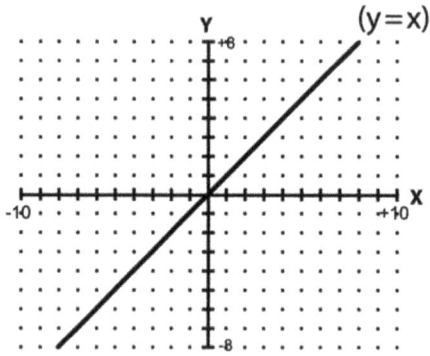

A. The steepness would change. Why or why not?

_____ NO   _____ YES   ____ STEEPER   ____ LESS STEEP

Fig. 1 Problem for Example 2

Fig. 2 Lines drawn by
students

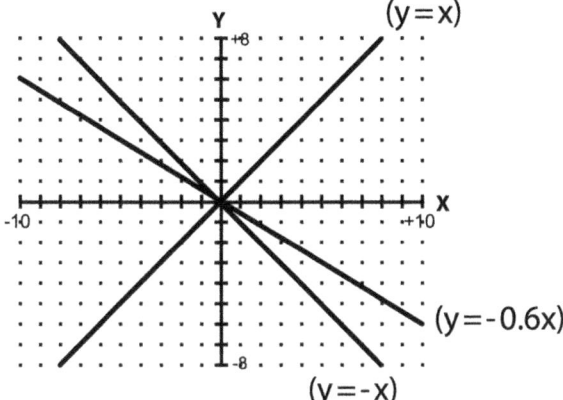

discussing whether this line was steeper than the line $y = x$.

Giselda proposed the second line was steeper and then decided it was less steep. Marcela repeatedly asked Giselda if she was sure. In the excerpt below, Marcela proposed that the line was less steep and explained her reasoning to Giselda. (Transcript annotations are in brackets; Translations are in italics beneath Spanish phrases.)

Marcela: No, it's less steeper...
Giselda: Why?
Marcela: See, it's closer to the x-axis...[looks at Giselda]...isn't it?
Giselda: Oh, so if it's right here...it's steeper, right?
Because look, let's say that this is the ground.

Entonces, si se acerca más, pues es menos steep
Then, if it gets closer, then it's less steep

...'cause see this one [referring to the line $y = x$]...is...

está entre el medio de la x y de la y. Right?
is between the x and the y.
Marcela: Porque fíjate, digamos que este es el suelo.
Giselda: [Nods in agreement.]
Marcela: This one [referring to the line $y = -0.6x$] is closer to the x than to the y, so this one [referring to the line $y = -0.6x$] is less steep.

Several strengths are evident in this discussion. First, the students combined multiple modes of communication, symbol systems, registers, and languages to communicate about a mathematical idea. Marcela coordinated several modes of communication—speaking and reading text, a graph, an equation. She coordinated two mathematical symbol systems, the graph (the line $y = x$, the axes) and the equations. She was reading, interpreting, and understanding not just the meaning of the

English text in the problem, but also reading, interpreting, and understanding the meaning of the equation and the lines on the graph.

Marcela also combined everyday and academic ways of talking to clarify the mathematical meaning of her description. She used two phrases typical of academic mathematical discourse: "Let's say" and the construction "If__, then__." Marcela used her everyday experiences and the metaphor that the x-axis is the ground ("Porque fíjate, digamos que este es el suelo" [*Because look, let's say that this is the ground*]). The everyday experience of climbing hills provided a resource for describing the steepness of lines (Moschkovich, 1996). Everyday meanings were strengths, not obstacles. Lastly, the students used two languages for their explanations and discussion, showing that both home and school languages are strengths for mathematical reasoning. Teachers must learn to notice how everyday language and experiences, including home languages, are, in fact, strengths for communicating mathematically.

We propose that teachers notice strengths by noticing the mathematical practices in what students say and do. Marcela's contributions reflect mathematical practices; she stated assumptions explicitly and connected her claims to two mathematical representations (graphs and equations). The phrase "If__, then__," reflects the practice of reasoning abstractly, and the phrase "Let's say this is__," reflects the practice of constructing arguments. She was also participating in the practice of paying attention to precision, by stating an assumption explicitly when she said, "Digamos que este es el suelo, entonces......" [*Let's say that this is the ground, then......*] (to decide whether a line is steeper or less steep, we first need a reference line for making this claim). She also connected a claim to the graph, another important mathematical practice. She supported her claim by making a connection to a mathematical representation; she used the graph, in particular the line $y = x$ and the axes, as references to support her claim that the second line was less steep. She used the axes as reference to support a claim about the line saying "Está entre el medio de la x y de la y" (*is between the* x *and the* y).

Opportunities for students to use strengths that are mathematical practices will depend on the quality and the activity structure of the tasks and policies enacted in classrooms. In this task, students needed to show conceptual understanding of slope, particularly when it is negative and less than 1. Explaining why the line would be steeper or less steep provided an opportunity for justifying one's reasoning. The activity structure required that students discuss their individual responses, arrive at a joint solution, and record that solution and explanation after reaching agreement (and before graphing the equation on the computer). Without this activity structure, the task might not reveal the student strength of mathematical practices such as constructing viable arguments and critiquing the reasoning of others. In this example, the students used Spanish and English without restrictions. Again, in contrast to policies that would restrict classroom talk to only the language of instruction, these students used home, school, and everyday languages to make sense of the mathematics.

3.3 Math Example 3: Leveraging Student Strengths and Expanding Classroom Math Practices

In this example, we examine a mathematics intervention (Rubel et al., 2016) and describe how it leveraged student strengths and expanded classroom mathematical practices to include community knowledge.

Using place-based pedagogy and critical mathematics approaches, a unit on statistics drew on the knowledge of Grade 12 students (17 and 18 years old), using the lottery as context. Students used and produced maps with digital tools to think critically (Rubel et al., 2016). The lottery provided a context that made the students' Funds of Knowledge (González et al., 2001), including linguistic resources, relevant. Students studied the lottery using maps that showed median income, total lottery spending, and net loss to the area under investigation (e.g., at neighborhood or state levels). They collected data and conducted interviews with community members. The unit supported statistical concepts, such as median, percentage, proportion, and inference. The unit went beyond procedures to support student engagement with STEM disciplinary practices such as modeling with mathematics (CCSS, 2010; NGSS, 2013), constructing arguments (CCSS, 2010), and arguing from evidence (NGSS, 2013).

The unit also supported students in participating in the mathematical practice of informal statistical inference (Makar & Rubin, 2018), related to modeling with mathematics (CCSS, 2010), and arguing from evidence (NGSS, 2013). Students combined their knowledge of the problem context (playing the lottery, characteristics of neighborhoods in their community) to use the data at hand as evidence to draw conclusions. The unit supported statistical literacy, requiring that students read data, find relationships within data, make claims beyond the data, and read behind the data to question its source (Morris, 2013; Shaughnessy, 2007). Students engaged in the "constant shuttling back and forth," (Pfannkuch, 2011, p. 29) between data and the real-world context, and the iterative cycle of creating and assessing conclusions required knowledge of the context (Wild & Pfannkuch, 1999).

The unit leveraged two student strengths in particular: street-level knowledge of the community, relevant to the construction and interpretation of maps and collecting data, and speaking Spanish, essential for conducting the interviews. Students made connections between data in the maps and their own experiences in those spaces. Knowledge of the context supported student engagement with the mathematical practice of modeling and the science practice of analyzing data. By engaging in statistical inference, the students also developed critical stances toward the lottery.

Spanish competency, another strength, supported data collection and interpretation. Using Spanish to conduct interviews allowed students to gather important interview data regarding lottery patronage from a sample representative of the community. Thus, they avoided excluding certain populations when gathering and interpreting data, allowing for more robust mathematical claims. Moreover, Spanish speakers were positioned as leaders because their linguistic resources were crucial for conducting interviews with monolingual community members. The unit design drew on students' strengths and the teaching practices assumed students brought such strengths. Students drew on these strengths to participate in disciplinary practices (constructing viable arguments, critiquing the reasoning of others, and using appropriate tools strategically) as they used maps and data to make arguments about the fairness of the lottery. The view of Spanish as a strength is particularly important because it eschews subtractive schooling policies and practices (Gibson, 1998; Valenzuela, 2005) that privilege assimilationist stances and would otherwise prohibit or denigrate its use.

This unit was based on the assumption that students bring mathematical and linguistic strengths such that their engagement with data collection, statistical analysis, and mathematical modeling would be productive. Furthermore, the unit expanded on what typically counts as mathematical practices in a classroom setting to include community knowledge. The success of this unit relied on students' everyday knowledge to make inferences from data. In this way, instruction honored students' knowledge of their own community as central to data collection and analysis.

4   Science Examples

There are varying and changing views about what counts as scientific thinking and practices. In their major review of scientific thinking research, Lehrer and Schauble (2015) argue that the current focus on science-as-practice, which became the basis for the policy document NGSS (2013), best captures the disciplinary practices of scientists and frames the most promising approach to policy in science education. Traditional approaches, often focusing on science as conceptual change (Lehrer & Schauble, 2015) focus on science as facts to be learned or processes (e.g., the scientific method) to be mastered. Similar to our analysis of mathematical examples, we use two science examples to illustrate our three recommendations: noticing students' strengths, recognizing science practices in student contributions, and expanding what counts as science practices. In particular, we focus on science practices emphasized by the NGSS (2013), including asking questions, analyzing and interpreting data, and arguing with evidence, as well as NGSS cross-cutting themes, specifically systemic thinking. These practices and themes overlap with scientists' disciplinary practices and are central to policy recommendations for classroom science instruction. We show that these practices and cross-cutting themes also reflect everyday cultural reasoning practices used by students from particular multilingual and non-dominant communities.

4.1 Science Example 1: Noticing Cultural Practices as Strengths, Recognizing Arguing as a Science Practice, and Expanding Science Practices

Unless policy makers and educators notice the strengths of children from marginalized communities, they may see them as underperforming. Hudicourt-Barnes (2003) rejects the idea that children from different communities should give identical responses when asked the same question. This expectation has led some researchers to paint a negative picture of the scientific abilities of Haitian immigrant students in the United States (Lee & Fradd, 1996; Lee et al., 1995), claiming that Haitian children's classroom strategies were inconsistent with the norms of science discourse. In contrast, Hudicourt-Barnes' work (2003) illustrates how to notice learners' strengths for learning, documenting Haitian children's classroom participation in sophisticated conversations about scientific phenomena using a conversational practice common in Haitian communities.

Haitian culture emphasizes spoken language for entertainment as well as communication. Adults and children participate in the social practice of *bay odyans* or *lodyans* which involves animated and entertaining interactions about a range of topics. These conversations take various forms, such as storytelling, reminiscing about previous experiences, and arguments (also called *diskisyon* or discussion) and occur in public settings, involving all members of the community (Hudicourt-Barnes, 2003). Usually, one person makes a claim and calmly defends it as one or more challengers question the claim, bring evidence, and engage the larger group. Other members of the group join in with laughter, approval, and other reactions. The goal is to entertain, but also to find the truth through argumentation.

Hudicourt-Barnes (2003) identified the social practice of bay odyans as a strength of Haitian students and recognized how it reflects argumentation using evidence in classroom science lessons, a key science practice. According to the NGSS, "As children move through the higher grades, they should participate more directly in comparison and critique of conflicting claims, including weighing respective strengths and weaknesses" (Lehrer & Schauble, 2015, p. 31).

In one observation of a group of Haitian students from a Grade 5/6 classroom (10 and 11 years old), students were documented expressing their arguments, evidence, and questions in a discussion about where mold would and would not grow (Hudicourt-Barnes, 2003). Children were asked to reflect on their life experiences, their previous learning, and their observations of mold growing on slices of bread in their classroom to inform their arguments. One child made a claim that mold grew easily in bathrooms. This prompted other children to engage with the idea, taking turns to provide evidence and questioning. Multiple children voiced their arguments and took on the role of challenger while the teacher acted as moderator, encouraging students to defend their positions. This example shows children providing explanations and evidence to support their perspectives by challenging one another using a familiar conversational pattern that is a strength for learning science. The example also provides evidence of their participation in the scientific practice of engaging with arguments using evidence (Hudicourt-Barnes, 2003).

In contrast with the question and known-answer format of traditional westernized classroom practices, the teacher from this classroom provided space for children to explore ideas using argumentation skills they developed in the practice of bay odyans (Hudicourt-Barnes, 2003). Because the teacher was aware of this cultural practice, they expanded what counts as a STEM practice beyond traditional expectations. This and intentional facilitation of a classroom discussion allowed children to engage more fully in the scientific practices than if they had followed westernized classroom dynamics (Hudicourt-Barnes, 2003). The student discussions included laughter and interjections, important elements in the practice of bay odyans. If the teacher in this classroom had held to a more rigid view of science practices, the rich student conversations may have been viewed as non-academic and shut down. The strengths children showed in the classroom discussion about mold mirror the authentic science practices of scientists and these practices need to be recognized in student discussions. When teachers provided opportunities for children to engage in bay odyans and employ their existing culturally relevant conversational practices during a science lesson, they were able to notice students' strengths (Hudicourt-Barnes, 2003), and recognize scientific practices. By investigating classroom discussions, researchers have shown that Haitian immigrant students' community practices reflect authentic scientific practices such as acquiring knowledge and searching for scientific meaning (Ballenger, 1997; Conant et al., 2001; Rosebery et al., 1992; Warren & Rosebery, 1995). This study also illustrates a more expansive and less culturally biased view that policy makers, researchers, and teachers can use to define what constitutes valid science practices (Hudicourt-Barnes, 2003).

4.2   Science Example 2: Recognizing Students' Strengths in Systemic Thinking and Expanding Science Practices

Considering what counts as science practices, Bang et al. (2012) discuss "settled expectations" (p. 303, citing Harris, 1995) in science and school that determine what are considered appropriate ways of talking, explaining, and understanding phenomena (Medin & Bang, 2014). For example, one biology practice involving categorizing objects and organisms as living versus nonliving fits an approach to science that prioritizes facts to be learned. Such settled expectations in science separate science from everyday experience, imposing on students what Bang et al. (2012) call the "nature-culture divide" (p. 303), preventing students from engaging with ideas at the boundary between their own experiences and the tenets of science. In line with our recommendation of expanding what counts as STEM practices, Bang et al. (2012) invite readers instead to "imagine the kinds of meaning-making that can arise within a desettling paradigm—that is one focused on…explicitly engaging students…at the nature-culture boundary." (p. 304).

Categorizing living versus nonliving asks students to use rigid definitions and learn the categories defined by scientists. This approach contrasts with the aims of the NGSS (2013), which include encouraging students to use systemic thinking. One of the cross-cutting concepts of the NGSS that can be applied across disciplines, "systems and system models," focuses on defining the boundaries of the system under study (National Research Council, 2012). Bang et al. (2012) argue that students' attempts to engage in "thinking at the edges," also referred to as "possibility thinking" are often not recognized in classroom activities focused on the more settled work of learning existing categories. They discuss an example reported by Warren and Rosebery (2011) where Jonathan, an African American male student in grade 7 (12 years old) questioned the sun's place in the category structure of living vs nonliving. Jonathan asked how the sun can be dead if it helps living things to live. A Euro-American student and the teacher responded that the sun cannot be thought of as a living thing. Jonathan eventually backed off, seemingly resigned that his point was misunderstood and that his view did not fit the system the teacher was using. However, Bang et al. (2012) point out that Jonathan was engaging in systemic thinking about the sun and how it relates to life. They connected Jonathan's thinking with how microbiologists think "at the edges" about microbial life forms, contesting existing boundaries and pushing the definition of "life" into new territory. Bang et al. (2012) use Helmreich's (2009) anthropological study of microbiologists' work to argue that active scientists' definitions of life are increasingly systemic and that human cultural experience and science are "more entangled than previously thought" (p. 307). Rather than assuming a deficit in Jonathan's ideas, this example illustrates how to notice the complexity of this student's thinking as a strength and recognize how he is engaging in a central science practice.

Bang et al. (2012) consider what desettling activities around nature-culture relations might look like using several classroom-related examples. We focus here on their final example of a design-based study of science learning for an urban indigenous community. Bang et al. (2012) discuss how in the initial design of this learning environment, the community-based team considered ways that indigenous knowledge systems relate to, as well as contrast with, Western science. One focus was the distinction between seeing humans as either "a part of" or "apart from" the natural world. This distinction between psychological distance versus closeness with nature is a theme in work comparing Native American with European American participants from the same rural area in the United States (Medin & Bang, 2014). For example, when asked how two animals and/or plants go together (Unsworth et al., 2012), Menominee children as young as 5 years were more likely than Euro-American children to mention ecological relations, such as linking the two species in the food chain ("the chipmunk would eat the berries") or mentioning that both have similar biological needs ("both need water to live"). Menominee children also more often justified the pairings using human closeness to nature, such as saying "I eat berries." Several other studies show similar examples of closeness to nature and ecological systems in Native American children and adults' thinking about biological species (Medin & Bang 2014). Marin and Bang (2018) reported yet another relational way of thinking about nature. In their investigation of Native American families' forest walks, they describe examples of observational practices such as reading land, as "a critical practice for *being* in the world as it enables relationship building with the natural world." (p. 92).

Noticing the strengths in systemic thinking that Native American youth bring to the classroom, and recognizing that these are, in fact, important science practices, the community-based design team emphasized relations among all things in nature (Bang et al., 2012). In focusing on river ecology, for example, they engaged students with activities in local settings, built on practices students had experienced (e.g., collecting edible and medicinal plants), and highlighted active relationships between organisms and habitats. In one case, they engaged students at an oxbow in a river—a place where changes over geological time can be noticed by reading land. When collecting water samples to assess the health of the river, teachers asked students to immerse themselves, wearing waist-high waders, and walking the river's earlier path. In contrast to the Western assumption of humans as dominant over nature, they presented humans in deference to plants and habitat (see also Bang et al., 2014). These activities made visible and supported the strengths of Native American students as systemic thinkers and provided opportunities for students to engage in science practices such as exploring boundaries, intersections, and dependencies across species.

This example illustrates noticing students' strengths, recognizing their links to science practices, and expanding the range of what are considered STEM practices. Bang et al. (2012) discuss ways that teachers and curriculum designers can assume students' strengths rather than deficits, creating opportunities for students to engage with scientific content and in science practices connected to their own lived experiences. Noticing these strengths and recognizing their links to science practices supports students in thinking like scientists by considering the system they are studying within a complex and interrelated context rather than engaging only with pre-differentiated chunks of information to be passively learned. Moving beyond settled definitions thus expands what counts as science practices.

5   Discussion

We see three important ways that research on multilingual and non-dominant students' strengths in mathematics and science education can inform policy and practices for STEM in multilingual settings. In this section, we review our three recommendations of noticing strengths, recognizing STEM practices in student contributions, and expanding what counts as STEM practices.

In the examples, we see important connections between science and mathematics policy standards and STEM practices relevant to instruction. Some practices—constructing arguments, using quantitative reasoning, and modeling—appear in both sets of standards and cut across disciplines. We note that asking questions is missing from current mathematical practice standards (CCSS, 2010) but is the first science practice (NGSS, 2013). This is puzzling given the extensive research on "problem posing" in mathematics education. Statistical inference is another practice that connects across mathematics and science, suggesting this may be an important cross-disciplinary practice for STEM instruction.

Research in mathematics and science education has resulted in policies recommending that instruction afford opportunities to participate in disciplinary practices to support students' STEM learning. As a start, multilingual and non-dominant learners need access to such opportunities. However, disciplinary practices need not be taught from scratch, some are already present in the practices of students' own communities or in students' contributions to classroom discussions, but often go unrecognized by instructors. We provided examples of how arguing with evidence and thinking about systems are science practices learners themselves bring to the classroom. We also illustrated how bilingual learners use their home language and everyday registers to communicate mathematically, thus making home and everyday ways of talking strengths students used to participate in mathematical practices (abstracting, generalizing, constructing arguments, and making claims more precise). We also showed that students' strength in local knowledge was central for collecting data, interpreting representations of that data, and making inferences.

Our first recommendation is to notice that learners bring strengths for doing and learning STEM. But noticing alone is not sufficient to create opportunities for students to participate in STEM practices. A necessary second step is to recognize disciplinary practices in what students say or do. The move is not only away from deficit views of multilingual learners and toward noticing students' strengths, but also to recognizing when and how those strengths reflect disciplinary knowledge and practices. In the words of Hudicourt-Barnes (2003, p. 76):

> We find that when Haitian children are in culturally familiar environments in classrooms focused on practicing science, the type of behavior they exhibit toward the acquisition of knowledge and the search for scientific meaning is deeply congruent with the practice of authentic scientific research.

Enacting policy that builds on student strengths requires considering how their strengths are relevant to classroom activity. The units and lessons described above leveraged students' linguistic strengths and community knowledge and supported student engagement with STEM disciplinary practices. However, home language practices and local knowledge can be strengths only if they are noticed and included. Activities were designed based on beliefs about students' strengths and engaging students in STEM practices. Local knowledge was a strength because it allowed students to engage in making sense. Language and life experiences helped children interpret data, bringing everyday and scientific practices together. In the science classroom, noticing and recognizing the similarity of argument structure in bay odyans with the argument structure in scientific reasoning led teachers to allow "everyday" argument as part of classroom work.

In this chapter, we have used an expansive view of student strengths and disciplinary practices. This unsettling perspective increases possibilities for students, so they use their own knowledge to contribute meaningfully to classroom work:

In our view, desettling entails imagining multivoiced meanings of core phenomena as open territory for sense-making in the science classroom, similar to the kinds of meaning-making opportunities that are available to scientists in the field. (Bang et al., 2012, p. 308)

Such a desettling perspective goes beyond traditional definitions of science and mathematics as separate. For example, as illustrated in the lottery unit, science and mathematics can come together and students can engage in practices from both disciplines. Using the three recommendations, policy and practice can embrace this desettling perspective and shift to treating STEM practices as one—but not the only—set of cultural practices (Medin & Bang, 2014) relevant to STEM learning. In particular, expanding what counts as STEM practices, policy and practice can recognize student contributions as perhaps different, but still scientifically and mathematically valuable and a foundation for further STEM learning.

References

Aguirre, J., Mayfield-Ingram, K., & Martin, D. (2013). *The impact of identity in K-8 mathematics: Rethinking equity-based practices*. Reston, VA: The National Council of Teachers of Mathematics.

Ballenger, C. (1997). Social identities, moral narratives, scientific argumentation: Science talk in a bilingual classroom. *Language and Education, 11,* 1–14.

Bang, M., Curley, L., Kessel, A., Marin, A., Suzukovich, E., & Strack, G. (2014). Muskrat theories, tobacco in the streets and living in Chicago as indigenous land. *Environmental Education Research, 20,* 37–55.

Bang, M., Warren, B., Rosebery, A. S., & Medin, D. (2012). Desettling expectations in science education. *Human Development, 55*(5–6), 302–318.

Barwell, R., Moschkovich, J., & Setati Phakeng, M. (2017). Language diversity and mathematics: Second language, bilingual, and multilingual learners. In J. Cai's (Ed.) *Compendium for research in mathematics education* (pp. 583–606). Reston, VA: NCTM.

Common Core State Standards. (2010). *Common core state standards for mathematical practice*. Washington, DC: National Governors Association Center for Best Practices and the Council of Chief State School Officers.

Conant, F., Rosebery, A., Warren, B., & Hudicourt-Barnes, J. (2001). The sound of drums. In E. McIntyre, A. Rosebery, & N. Gonzalez (Eds.), *Building bridges: Linking home and school* (pp. 51–60). Portsmouth, NH: Heinemann.

Faltis, C. J., & Valdés, G. (2016). Preparing teachers for teaching in and advocating for linguistically diverse classrooms: A vade mecum for teacher educators. In D. H. Gitomer & C. A. Bell (Eds.), *Handbook of research on teaching* (5th ed., pp. 549–592). Washington, DC: American Educational Research Association.

Gholson, M., Bullock, E., & Alexander, N. (2012). On the brilliance of black children: A response to a clarion call. *Journal of Urban Mathematics Education, 5*(1), 1–7.

Gibson, M. A. (1998). Promoting academic success among immigrant students: Is acculturation the issue? *Educational Policy, 12*(6), 615–633.

González, N., Andrade, R., Civil, M., & Moll, L. (2001). Bridging funds of distributed knowledge: Creating zones of practices in mathematics. *Journal of Education for Students Placed at Risk, 6*(1–2), 115–132.

Goodwin, C. (1994). Professional vision. *American Anthropologist, 96*(3), 606–633.

Gutiérrez, K. (2008). Developing sociocritical literacy in the third space. *Reading Research Quarterly, 43*(2), 148–164.

Gutiérrez, K. D., & Rogoff, B. (2003). Cultural ways of learning: Individual traits or repertoires of practice. *Educational Researcher, 32*(5), 19–25.

Hammer, D. (1996). More than misconceptions: Multiple perspectives on student knowledge and reasoning, and an appropriate role for education research. *American Journal of Physics, 64*(10), 1316–1325.

Harris, C. I. (1995). Whiteness as property. In K. Crenshaw, N. Gotanda, G. Peller, & K. Thomas (Eds.), *Critical race theory* (pp. 276–291). New York: New Press.

Helmreich, S. (2009). *Alien ocean: Anthropological voyages in microbial seas.* University of California Press.

Hudicourt-Barnes, J. (2003). The use of argumentation in Haitian Creole science classrooms. *Harvard Educational Review, 73*(1), 73–93.

Kilpatrick, J., Swafford, J., & Findell, B. (2001). Adding it up: Helping children learn mathematics (Vol. 2101). National research council (Ed.). Washington, DC: National Academy Press.

Lee, O., & Fradd, S. (1996). Interactional patterns of linguistically diverse students and teachers: Insights for promoting science learning. *Linguistics and Education, 8,* 269–297.

Lee, O., Fradd, S., & Sutman, F. (1995). Science knowledge and cognitive strategy use among culturally and linguistically diverse students. *Journal of Research in Science Teaching, 32,* 797–816.

Lehrer, R., & Schauble, L. (2015). The development of scientific thinking. In R. M. Lerner (Ed.), *Handbook of child psychology and developmental science* (7th ed.). John Wiley & Sons.

Louie, N. L. (2018). Culture and ideology in mathematics teacher noticing. *Educational Studies in Mathematics, 97*(1), 55–69.

Makar, K., & Rubin, A. (2018). Learning about statistical inference. In D. Ben-Zvi, J. Garfield, & K. Makar (Eds.), *International handbook of research in statistics education* (pp. 261–294). Switzerland: Springer.

Marin, A., & Bang, M. (2018). "Look it, this is how you know:" Family forest walks as a context for knowledge-building about the natural world. *Cognition and Instruction, 36,* 89–118.

Martin, D. B. (2019). Equity, inclusion, and antiblackness in mathematics education. *Race Ethnicity and Education, 22*(4), 459–478.

Martínez, R. A., & Mejía, A. F. (2020). Looking closely and listening carefully: A sociocultural approach to understanding the complexity of Latina/o/x students' everyday language. *Theory into Practice, 59*(1), 53–63.

Mason, J. (2002). *Researching your own practice: The discipline of noticing.* Psychology Press.

Medin, D. L., & Bang, M. (2014). *Who's asking? Native science, western science, and science education.* Cambridge, MA: MIT Press.

Morris, K. (2013). *Map reading framework.* Unpublished manuscript, Department of Teaching, Learning, Policy, & Leadership, University of Maryland, College Park, MD.

Moschkovich, J. N. (1996). Moving up and getting steeper: Negotiating shared descriptions of linear graphs. *The Journal of the Learning Sciences, 5*(3), 239–277.

Moschkovich, J. N. (1998). Resources for refining conceptions: Case studies in the domain of linear functions. *The Journal of the Learning Sciences, 7*(2), 209–237.

Moschkovich, J. N. (1999a). Students' use of the x-intercept as an instance of a transitional conception. *Educational Studies in Mathematics, 37,* 169–197.

Moschkovich, J. N. (1999b). Supporting the participation of English language learners in mathematical discussions. *For the Learning of Mathematics, 19*(1), 11–19.

Moschkovich, J. N. (2002). A situated and sociocultural perspective on bilingual mathematics learners. *Mathematical Thinking and Learning, 4*(2&3), 189–212.

Moschkovich, J. N. (2011). Supporting mathematical reasoning and sense making for English Learners. In M. Strutchens & J. Quander (Eds.), *Focus in high school mathematics: fostering reasoning and sense making for all students* (pp. 17–36). NCTM: Reston, VA.

Moschkovich, J. N. (2013). Principles and guidelines for equitable mathematics teaching practices and materials for English language learners. *Journal of Urban Mathematics Education, 6*(1), 45–57.

Moschkovich, J. N. (2015). Scaffolding mathematical practices. *ZDM, The International Journal on Mathematics Education, 47*(7), 1067–1078.

National Research Council. (2012). *A framework for K-12 science education: practices, crosscutting concepts, and core ideas.* Washington, DC: The National Academies Press.

National Academies of Sciences, Engineering, and Medicine. (2018). *English learners in STEM subjects: Transforming classrooms, schools, and lives.* Washington, DC: The National Academies Press.

NGSS Lead States. (2013). *Next generation science standards: For states, by states.* Washington, DC: The National Academies Press.

O'Connor, C., & Michaels, S. (1993). Aligning academic task and participation status through revoicing: analysis of a classroom discourse strategy. *Anthropology and Education Quarterly, 24*(4), 318–335.

Pfannkuch, M. (2011). The role of context in developing informal statistical inferential reasoning: A classroom study. *Mathematical Thinking and Learning, 13*(1–2), 27–46.

Rosebery, A., Warren, B., & Conant, F. (1992). Appropriating scientific discourse: Findings from language minority classrooms. *Journal of the Learning Sciences, 2*(1), 61–94.

Rubel, L. H., Lim, V. Y., Hall-Wieckert, M., & Sullivan, M. (2016). Teaching mathematics for spatial justice: An investigation of the lottery. *Cognition and Instruction, 34*(1), 1–26.

Schoenfeld, A. H. (1992). Learning to think mathematically: Problem solving, metacognition, and sense-making in mathematics. In D. Grouws (Ed.), *Handbook for research on mathematics teaching and learning* (pp. 334–370). New York, NY: Macmillan.

Shaughnessy, J. M. (2007). Research on statistics learning and reasoning. In F. Lester's (Ed.), *Second handbook of research on mathematics teaching and learning* (pp. 957–1011). Reston, VA: National Council of Teachers of Mathematics.

Smith, J. P., III, diSessa, A. A., & Roschelle, J. (1994). Misconceptions reconceived: A constructivist analysis of knowledge in transition. *The Journal of the Learning Sciences, 3*(2), 115–163.

Unsworth, S. J., Levin, W., Bang, M., Washinawatok, K., Waxman, S. R., & Medin, D. L. (2012). Cultural differences in children's ecological reasoning and psychological closeness to nature: Evidence from Menominee and European American children. *Journal of Cognition and Culture, 12*, 17–29.

Valdes-Fallis, G. (1978). Language in education: Theory and practice: Vol. 4. In *Code switching and the classroom teacher.* Wellington, VA: Center for Applied Linguistics.

Valenzuela, A. (2005). Subtractive schooling, caring relations, and social capital in the schooling of US-Mexican youth. In *Beyond silenced voices: Class, race, and gender in United States schools* (pp. 83–94). NY: SUNY Press.

Warren, B., Ballenger, C., Ogonowski, M., Rosebery, A. S., & Hudicourt-Barnes, J. (2001). Rethinking diversity in learning science: The logic of everyday sense-making. *Journal of Research in Science Teaching: The Official Journal of the National Association for Research in Science Teaching, 38*(5), 529–552.

Warren, B., & Rosebery, A. (1995). Equity in the future tense: Redefining relationships among teachers, students, and science in linguistic minority classrooms. In W. Secada, E. Fennema, & L. Adajian (Eds.), *New directions in mathematics education* (pp. 298–328). New York: Cambridge University Press.

Warren, B., & Rosebery, A. S. (2011). Navigating interculturality: African American male students and the science classroom. *Journal of African American Males in Education, 2*(1), 99–115.

Watson, A. (2009). Thinking mathematically, disciplined noticing and structures of attention. In *Mathematical action & structures of noticing* (pp. 211–222). Brill Sense.

Wild, C. J., & Pfannkuch, M. (1999). Statistical thinking in empirical enquiry. *International Statistical Review, 67*(3), 223–248.

Salvador Huitzilopochtli is a Ph.D. Candidate in Mathematics Education at the University of California, Santa Cruz. His research interests include mathematical argumentative writing and early algebra with a focus on creating instructional environments that foster equitable outcomes for all students. He earned Master's degrees in Education from UC Berkeley and UC Santa Cruz. Mr. Huitzilopochtli's research is informed by ten years of experience as a middle-school mathematics teacher and teacher leader in culturally, linguistically, and economically diverse schools in the San Francisco East Bay Area. Mr. Huitzilopochtli's work centers equity and draws upon his additional experience working in violence prevention, academic support, cultural education, and mentoring.

Julianne Foxworthy Gonzalez is a Ph.D. candidate in Mathematics Education at the University of California, Santa Cruz. Her research interests include mathematical discussions, statistics education, and equitable instructional practices. Before pursuing educational research, she taught fifth and sixth grade in elementary school in culturally, linguistically, and economically diverse contexts for six years.

Judit N. Moschkovich is a Professor of Mathematics Education in the Education Department at the University of California, Santa Cruz. Her research uses sociocultural approaches to study mathematical thinking and learning, algebraic thinking, mathematical discourse, and mathematics learners who are bilingual, learning English, and/or Latino/a. Her work has appeared in the *Journal for Research in Mathematics Education, Educational Studies in Mathematics, the Journal of Mathematical Behavior, the Journal of the Learning Sciences, and Cognition & Instruction.* She edited *Language and mathematics education: Multiple perspectives and directions for research* (2010) and coedited *Mathematics education and language diversity* (2016). She served on the Consensus Committee for "English Learners in STEM subjects: Transforming classrooms, schools, and lives: A report from the National Academies of Science, Engineering, and Medicine" (2018). She is a 2018 Fellow of the American Educational Research Association (AERA) and received the 2019 Distinguished Scholar Award, Special Interest Group Research in Mathematics Education (SIG-RME) in AERA.

Sam R. McHugh is a Ph.D. student in Developmental Psychology at the University of California, Santa Cruz. Their research focuses on how young children make sense of the world from everyday interactions with parents. They received a Master's in Psychology in 2019 and co-authored a Monograph of the Society for Research in Child Development ("Exploration, Explanation, and Parent-Child Interaction in Museums"), published in 2020.

Maureen A. Callanan is a Professor of Developmental Psychology at the University of California, Santa Cruz. Her research focuses on young children's developing understanding of the natural world in the context of family conversations. She takes a sociocultural approach, investigating young children's language and cognition with attention to diversity across families and communities. She has a long-standing research partnership with Children's Discovery Museum of San Jose, where she has been PI or co-PI on several National Science Foundation funded projects investigating children's and families' informal learning about science. She was a faculty partner in the Center for Informal Learning and Schools (CILS), and a coauthor of the 2009 National Academies Press volume: *Learning Science in Informal Environments:People, Places and Pursuits.* She has been an Associate Editor for *Psychological Bulletin* and *Journal of Cognition and Development*, past chair of AERA's SIG focused on Informal Learning Environments Research, and is President-Elect of the Cognitive Development Society.

Multilingual Students Working with Illustrated Mathematical Word Problems as Social Praxis

Laura Caligari, Eva Norén, and Paola Valero

Abstract Word problems in mathematics can present challenges for multilingual students. Previous research shows that the language and cultural contexts of word problems are major obstacles for many students. This study advances the research by considering the work with illustrated word problems as a social praxis. We specifically ask what social, cultural and linguistic experiences multilingual students mobilise and create when working with illustrated word problems. Data were collected from eight multilingual students in fourth grade. The students were given word problems to solve individually. The data were qualitatively analysed based on a four-fold structure of social praxis as a framework. Findings reveal that word problems will always mobilise exclusion. The study concludes that when working with illustrated word problems teachers need to recognise this condition and balance the ways of working with word problems. Policies need not only to promote linguistic and cultural diversity explicitly. They could open real possibilities for policy enactment where the linguistic and cultural differences embedded in illustrated mathematical word problems are discussed and negotiated.

Keywords Illustrated mathematical word problems · Multilingual students · Social praxis

L. Caligari (✉) · E. Norén · P. Valero
Department of Mathematics and Science Education, Stockholm
University, Stockholm, Sweden
e-mail: laura.caligari@mnd.su.se

E. Norén
e-mail: eva.noren@mnd.su.se

P. Valero
e-mail: paola.valero@mnd.su.se

1 Introduction

Educational policy in Sweden has attended to research results that, since the 1990s, show the benefit of including and supporting students' mother tongue as a resource for instruction and learning in mathematics (Barwell, 2009; Moschkovich, 2007; Prediger & Schueler-Meyer, 2017). According to the Swedish national curriculum, all teaching must be adapted to each student's prerequisites and needs. Teaching should promote students' continued learning and knowledge development based on their backgrounds, past experiences, languages and knowledge (Skolverket, 2011). Indeed, all students with first languages other than Swedish have the right to tuition in their first language. "Mother tongue" is a subject in the Swedish curriculum, thus, students living in municipalities that have the resources and organise for the subject will have the chance of studying their first language in school. Studies in Swedish mathematics classrooms (Norén, 2010; Ryan, 2019) show that multilingual students benefit when they are able to utilise their linguistic resources and cultural identities.

However, for multilingual students and in multilingual classrooms—where students speak at least two languages and where more than one language could be used in the classroom (Barwell, 2009, 2018)—the fact that Swedish is the instructional language poses challenges. This results in the systematic underperformance of multilingual students in mathematics compared to their first-language Swedish peers (Skolverket, 2019a, 2019b). One of such challenges is how students' linguistic and cultural knowledge and experience are significantly taken as a resource. In fact, the last quality review by the School Inspectorate revealed a lack of knowledge about multilingual students' cultural backgrounds and experiences in many schools (Skolinspektionen, 2010). Furthermore, research has shown that there is a lack of understanding of multilingual students' experiences in multilingual mathematics classrooms and how they can become significant for mathematics (Svensson Källberg, 2018). In other words, the Swedish educational policy has a clear intention of inclusion but keeps on producing a clear exclusion of multilingual students, particularly in mathematics. This is problematic, in terms of equity, for students learning mathematics in a second language (Barwell et al., 2019).

In this chapter, we explore this challenge as it unfolds in one important element of school mathematics: the reading and solving of illustrated mathematical word problems. In the Swedish mathematics curriculum, problem-solving is a privileged competence that students must develop (Skolverket, 2011). It is emphasised that problems should relate to students' everyday lives (Skolverket, 2017) as a way to bridge school mathematics and out of school reality. Despite their centrality in many curricula, it is known that word problems are difficult for many students, in particular for multilingual ones (Barwell, 2009; Cooper & Dunne, 2000). If word problems are defined as "one way to express beliefs about how everyday experience and mathematics should be related in order for math learning to take place effectively" (Lave, 1992, p. 75), the connections between the beliefs about everyday experiences encapsulated in word problems and the actual experiences of students are a clear predicament for research and practice. Existing research has shown that, when solving word problems about everyday life, multilingual students tend to draw on their first language (Planas & Civil, 2013) and on their cultural or religious identities (Barwell, 2009) to increase their understanding and learning. But to maintain and continue developing, it is important that the experiences they create and draw on are valued and utilised. This is hardly the case.

This chapter advances the research by considering the work with illustrated word problems as a social praxis, in which word problems and students' experiences intertwine as "learning and knowledge [become] situated in social interaction" (Gutiérrez, 2013, p. 45). The overall aim of this study is to explore what experiences multilingual students draw on and create when solving illustrated mathematical word problems and to discuss the implications for mathematics classrooms. This study more specifically addresses the question of what social, cultural and linguistic experiences do multilingual students mobilise and create when engaging in the social praxis of solving illustrated mathematical word problems in Swedish. Based on our findings, we provide some insights for policy.

2 Mathematical Word Problems and Multilingual Students

Mathematical word problems have existed at least since Babylonian times, as a pedagogical tool to induce mathematical activity. Working with mathematical word problems is nowadays regarded as an effective way for students to learn and become successful in mathematics (Boonen et al., 2016; Dyrvold et al., 2015). Word problems combine the cues of mathematics within a context in which the mathematics is to be used. Thus, they often reflect general characteristics of a given society (Barwell, 2018; Gerofsky, 1996).

Research shows that students' engagement with the contexts in word problems is challenging. Some aspects of the context are often ignored in students' solutions, and seemingly "realistic" items may cause confusion and misunderstanding, leading to nonsensical answers (De Corte et al., 2000; Greer, 1993, 1997). It can be challenging for native speakers and multilingual students to communicate mathematical ideas and concepts in relation to word problems (Barwell, 2009; Clarkson, 2007). This has raised the question of whether word problems may be too artificial or the contexts too unfamiliar for students. Explorations of such difficulties have revealed that students engage in considering the contexts in word problems. Multilingual students often relate to their own cultural experiences and home cultures when solving and constructing word problems of their own (Barwell, 2009). Barwell (2018) points that students' solutions only become nonsensical when the solutions lack students' interactive processes and the sense-making that led up to them. He stresses that word problems must be understood as "socially constructed, deployed and interpreted texts" (p. 102). They are social texts, and as such reflect a tension between normative academic practice and students' life experiences (Barwell, 2018).

As an important element of a normative academic practice, mathematical word problems constitute a genre of mathematical literacy, with specific linguistic, structural and contextual features that students need to familiarise themselves with (Barwell, 2009; Gerofsky, 1996, 2004). Students' knowledge of the strategies and routines in the mathematical literacy genre are therefore significant. The word problem genre has certain textual forms but can also be regarded as social (Barwell, 2018). Students need to understand the norm—how they are expected to read, interpret, respond—so it reflects their mathematical knowledge. Students from minority language/cultural groups and students from working-class backgrounds often perform less well on word problems than their counterparts (Cooper & Dunne, 2000). In Barwell's (2018) words, word problems can create social stratification between those who are "at home" with the genre and those who are not. Moreover, reading a mathematical word problem in a second language often entails putting more effort into decoding the text and understanding the context than interpreting the mathematical content (Clarkson, 2009). Thus, performance on word problems is partly related to students' knowledge, while stratification of students into groups happens according to their backgrounds.

3   Mathematical Word Problems and Illustrations

Word problems may also include illustrations. Word problems can have a textual and a graphic component, and in some cases these components are more or less related and significant for engaging with the mathematical task. Illustrations are pedagogically intended to help learners visualise the context and thus solve the word problems more easily (Dewolf et al., 2015). Dewolf et al. (2015) explain that the idea with representational illustrations is to help students construct a rich mental model of the mathematical situation and prevent them from only searching for a standard computation of the word problem. Teledahl and Olsson (2021) suggest that students engage written and illustrated word problems in various ways, as an illustration and a written text in a problem can be viewed as two different sources of information and can be treated as isolated, connected or combined. These studies have shown that, despite the good pedagogical intentions and the different types of engagement, students tend to neglect the representational illustrations when solving the problems.

The question emerges, of whether this is also the case when multilingual students meet illustrated word problems. Besides facilitating a cognitive demand, illustrations can be seen as a way of generating a familiar connection between the students and the mathematical task. Research on textbooks and instructional materials points to the fact that mathematical contents are presented together with national cultural elements such as forms of behaviour, artefacts, geography, identities and history (Fan et al., 2018). Doğan and Haser (2014), in a study of Turkish textbooks, have pointed to the potential effects of exclusion produced when students from cultural minorities meet problems that emphasise a particular national identity. Souza and da Silva (2018) have also shown how the use of toys in the problems and illustrations in Brazilian primary mathematics textbooks offer a traditional gender role for girls to relate to. In other words, word problems in curricular materials incorporate elements of the national culture and have effects on how students from particular groups engage them.

In short, multilingual students mobilise their knowledge and experiences of the world out of school when meeting word problems with their texts and illustrations. The text and illustrations in word problems carry with them particular cultural elements. Then one can expect many issues emerging in the interactions between students and illustrated word problems; and such issues go beyond the linguistic and cognitive capabilities required to successfully engage with the problems. To explore this issue, we propose to move the conceptualizations of illustrated word problems as social texts—as proposed by Barwell (2018)—and cognitive devices—as proposed by Dewolf et al. (2015)—into the terrain of social praxis.

4 Working with Illustrated Word Problems as Social Praxis

Radford's cultural theory of learning (2008a, 2008b, 2018) allows us to understand the work of students solving illustrated word problems as a social praxis, that is, a process where social and cultural forms of knowing are constituted. Such work can be conceived not as an interaction between two independent entities—the student and the problem—but as an entanglement between the student and the illustrated word problems as cultural artefacts, where learning and becoming emerges. The activity of working with problems binds in inseparable ways who the student is—her experiences and ways of making sense mathematically—and the cultural significations that are encapsulated in the illustrated word problem. In this way, the engagement of the student with the problem is not just the mediation of her thinking for the purpose of objectifying the mathematical knowledge in the problem, but also a meeting with the cultural significations that are part of the contexts to which the words and illustrations in the problem refer to.

Illustrated word problems instantiate ideas of society, and embody cultural perceptions about mathematics, the world and individuals. The cultural environment provides illustrated word problems with "raw material" and regulates what is being perceived as norm in a society, reinforcing cultural perceptions about the world and individuals (Radford, 2018). Thus, word problems understood as artefacts facilitate the assimilation of perceptions not only of mathematics—their technical aspect—but also of the very same culture they encapsulate—their cultural aspect. This is what Radford highlights with the idea that artefacts are given meaning with respect to cultural systems of signification (e.g. Radford, 2018, p. 454).

Students are expected to identify with situations in the problems and understand them since they are supposed to be part of the culture shaping illustrations and word problems in curricular materials. However, multilingual students' subjective meanings may clash with the cultural objective meaning in illustrated word problems, since their culture, language and experiences may differ from those of the dominant

Semiotic System of Cultural Signification

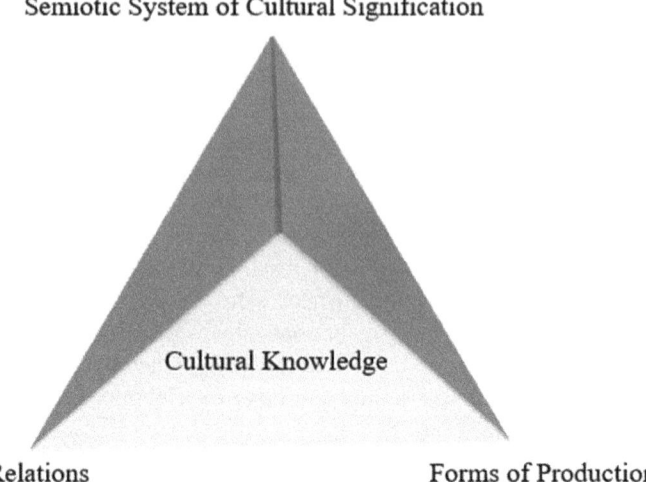

Cultural Knowledge

Forms of Social Relations   Forms of Production

Fig. 1 The four-fold structure of social praxis (Radford & Empey, 2007, p. 235)

culture encapsulated in problems in textbooks for school instruction. Radford and Empey's (2007) four-fold structure of social praxis (see Fig. 1) provides a framework to examine multilingual students' work with illustrated word problems and explore how their experiences are mobilised in that social praxis.

The first element is *Forms of Social Relations,* such as interactions, or divisions of labour. In this study, this element implies the interaction between the student and the illustrated word problem as well as statements about previously experienced interactions when working with illustrated word problems. Division of labour refers to assumed or assigned roles, which are often more like characters, that differentiates the self and the other. We explore the roles that students have experienced previously working with word problems.

The second element is *Forms of Production*—the artefacts that mediate cultural perceptions like signs and objects, in this study texts and illustrations. Students' interactions with the texts and illustrations can help us explore their experiences, articulated in an interactive process when working with the problems.

The third element is the *Semiotic System of Cultural Signification,* for instance, a cultural perception about the world and individuals. Students expressed perceptions that have a normative function, such as what is good, right, the truth, methods of inquiry or the legitimate forms of knowledge representation relating to working with illustrated word problems. This element also helps us explore how students perceive themselves, experience culture and inherent perceptions.

The fourth element is *Cultural Knowledge,* which is an epistemological point of view, and involves the knowledge available in a culture. This refers to a process that includes the three former elements. In this study, the fourth element relates to unspoken rules relating to activity, language and norms students reveal when working with illustrated mathematical word problems.

5   Methodology

This qualitative study explores the entanglement between multilingual students and illustrated word problems as cultural artefacts, as they work with them. Our focus is not on students' right or wrong solutions to the problems, but on their experiences.

5.1   About the Participating Students

We asked a school in a socio-economically disadvantaged area outside a major Swedish city if some of their students wanted to participate in our study. We chose to do our study in fourth grade because in Sweden the transition from third- to fourth-grade entails moving from lower to middle school, changing teachers, engaging new mathematics textbooks and more reading. We also expected that students would have experienced working with illustrated world problems due to their centrality in the curriculum. We informed the fourth graders' parents about the study and ethical issues, and those interested signed a consent form. Eight multilingual students agreed to participate. After their mother tongue, Swedish is their second language and English a third language since the latter is compulsory at least from third grade. The eight students had all started attending this school in first grade. In their class of 25 students, all had Swedish as their second language and 12 different mother tongues were represented in the classroom. Swedish was the instructional language.

On the day of the interviews, students were informed individually about the study and the procedure, and they gave verbal consent before starting. The interviews and engagement with the word problems were conducted individually by the first author in a room next to the students' classrooms. No time limit was imposed, and each encounter lasted about 20 min. All students' names are pseudonyms.

To begin with, the students got to answer some interview questions, like, "*what do you learn when working with word problems?*", "*would you prefer to have word problems in your first language?*" and "*how do you work with word problems in class?*" Some questions needed follow-ups, like, "*explain a little more about [...]*" when the student had answered too briefly. The aim with the questions was to get background information about the students, talk about their experiences and thoughts related to word problems, and help them feel comfortable with the interview situation. We did not ask questions about their mathematics scores, mathematics abilities or language proficiency; we wanted to focus on their experiences. Even if we noticed that when students spoke Swedish, they seemed fluent, we had no idea of their level of vocabulary and competence in Swedish.

After the interview questions, the students were given the word problems, blank paper, a pencil and an eraser. We asked the students to read the word problem out loud and to think aloud while working. When they fell silent, they were asked, "*what are you thinking now?*". We did not specify whether they should think aloud in their

164. How much older is a) Daddy than mummy? b) Jonas than Miranda?
165. Look at the map. The journey started in Ljungby. The first day they went to Jönköping and then to Mjölby. How far was it?

164. Hur mycket äldre är a) pappa än mamma? b) Jonas än Miranda?
165. Titta på kartan. Resan började I Ljungby. Första dagen åkte de till Jönköping och sedan till Mjölby. Hur lång *sträcka* blev det?

Fig. 2 This is the family Svensson from Ljungby in Småland. Last summer they went to Kolmården zoo (Undvall et al., 2011, p. 43; illustration by Unenge [Johan Unenge has given his permission to use his illustrations in this text.])

first or second language, but to proceed as they normally would when solving a word problem.

5.2 About the Tasks

The word problems were chosen based on our knowledge that these problems had troubled fourth-grade multilingual students at another school (Norén & Caligari, 2021). The word problems were selected from a year four mathematics textbook (Undvall et al., 2011) that was not used in the participants' school, so they were not familiar with the book or the problems. The problems include informational illustrations (Elia & Philippou, 2004) containing numbers not provided in the text but essential for solving the problem. For example, the illustrations include prices of items, ages of people and distances.

Two of the tasks had the theme *Kolmården* (see Fig. 2). Kolmården is a huge zoo with animals from around the world. Two other tasks had the theme *The Market* (see Fig. 6).

5.3 About the Analysis

The interviews were audio recorded, stored and transcribed. The data consisted of transcripts and participants' solutions produced during the interviews. The analysis began with a close reading of the transcripts, which were then searched for illustrative passages relating to the categories in the analytical framework. We analysed the students' statements and the work processes and sense-making that lead them to their solutions, and finally highlighted social, cultural and linguistic experiences. Table 1 indicates the kind of connections that we established between the framework

Table 1 Analytical framework

Category	Sub-category	Sub-category	Identification rule(s)
Forms of social relations	Interactions Roles	• teacher • student • word problem • themselves • others	Expressed interactions between the student and teacher, student and other students, student and word problem Expressed roles students give themselves and/or others
Forms of production	Artefacts	• text and illustration	How the students read and produce texts and illustrations
Semiotic system of cultural signification	Perceptions	• ideas • methods • oneself	Expressed perceptions relating to word problems. For example, good, bad, difficult. Expressed perceptions of method when working on word problems Expressed perceptions of oneself
Cultural knowledge	Unspoken rules	• activity • language • norms	Expressed demonstrated understanding of unspoken rules when working with word problems, words and illustrations

and evidence emerging in the interviews and conversations around the work with the problems.

6 Findings

6.1 Forms of Social Relations

In the interviews, the participants said that on occasion, when a word problem is too difficult, they work in smaller groups in an adjoining room and that "*the teacher explains difficult words*". But usually, they worked individually with word problems in the classroom. One participant explained that it should be "*quiet so you don't disturb each other*". The participants' interactions with the word problems were evident while they read and looked at the task, and some students said things like, "[*d*]*o I need to count this?*" or "[*w*]*ait, what?*" or "[*i*]*s this a task?*" And sometimes while writing a solution, "[*h*]*ow do I write this?*" The interaction seemed to be with the word problem, or with themselves.

One of the roles highlighted by students was the teacher–student role. The student's role implies learning and the teacher is assumed to be the only one who knows which words to practice and learn; in other words, "if we don't understand we can ask the teacher for tasks to practice at home".

The idea of the other is related to contrast and evaluation. The other is a fellow student to compare oneself with. Either the other is defined as the one having difficulties when working with word problems or the one who thinks they are easy: "*They* [word problems] *may be too difficult for some, but I think they are fun*", or the opposite: "*The others think they* [word problems] *are easy, but I don't*".

6.2 Forms of Production

As the participants read the word problems out loud it appeared clear that some students switched their readings between the text and illustration, and some of the words seemed to be difficult to read and understand. The participants got stuck and it showed as they slowed down their reading, sounded the words or reread them. "*This is the family Svensson from Lj… Lju… Ljung… by in Små… land… in Småland. Last… summer, they went to Kolo… Kolmården… zoo. How much older is… Peter forty-two years, Anna thirty-nine years, … Mar… Miranda three years, O… Olivia ten years, Jonis twelve years. How much older is daddy than mummy, Jonis than… Miranda?*" (see Fig. 2). The most challenging words to read were frequently (perhaps unfamiliar) names of cities such as *Ljungby, Småland* and *Jönköping* and names of people such as *Patrik, Miranda, Olivia* and *Jonas*. One participant reasoned "*Miranda? Miranda is a drink in Morocco*". While three participants read "[How much older is] *Jonis than Miranda*" instead of Jonas, as written in the word problem. Jonis is a common Arabic name.

This challenge while reading is something that needs to be taken into consideration as it can affect students when they are working on word problems with time constraints.

In task 165 (see Fig. 2) the illustration of the map carries meaning; that is, it provides information critical to the word problem. Not understanding the map is thus a disadvantage. Likewise, misreading or not understanding how to interpret and respond to a question, like, "how far was it?", can result in the student not demonstrating the required mathematical knowledge. As one participant reasoned out loud while writing an answer: "*How long was the line? It was long! They walked all this way to Kolmården zoo*" (see Fig. 3).

Fig. 3 Student C solution to
task165 (see Fig. 2)

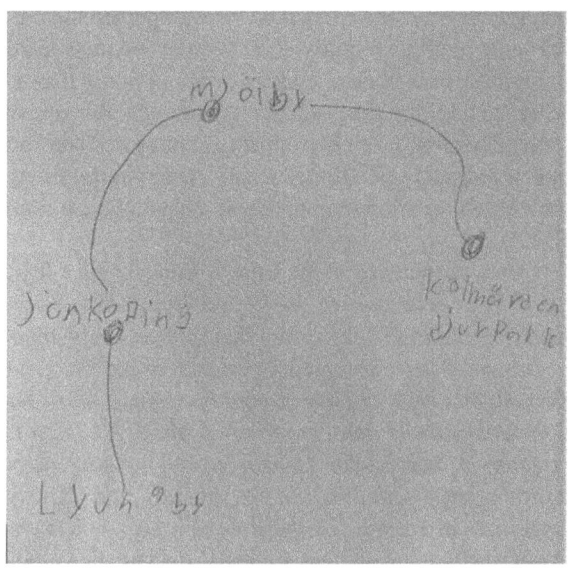

6.3   Semiotic System of Cultural Signification

The students said that the main idea of working with word problems is to "... *learn how to count in different ways*" or "*it teaches me different difficult things, like multiplication and division*". Also, there was the perception that word problems are more complicated to work with than mathematical tasks without words. During the interviews, students said word problems were more difficult. If the word problem was considered difficult (by the teacher or the students), they usually read the text with the teacher and focused on difficult words—"*it's good because you get to have more explanations*".

The participants said that when working with word problems the most suitable method is reading and understanding what to do next. They did not refer to illustrations in their statements: "*it is difficult when it is a text, you have to read carefully*"; "*In math, it is a little harder to figure out the text. The hard thing is that you have to understand the text. I don't really understand the text all the time*"; "*you have to read and then you must count*". Before writing their solutions, students sometimes ask themselves "*what should I write?*". The students here are seeking legitimate forms of knowledge representation, which shows that they understand they must produce a text of some kind, also showing familiarity with the word problem genre (Barwell, 2009). Since the students were talking aloud while working with the problems, their interactive process and the meaning making that led to a solution were evident. This revealed, among other things, that some students first verbalised an answer before writing it down. In Task 164a (see Fig. 2), participants A, E and D counted aloud while raising a finger at each accentuated word, then looked at the fingers and said an answer; that is, they said, "*thirty-nine, **forty**, **forty-one**, **forty-two**... three*", three being their answer as they held three fingers in the air. Despite the fact they had used the same method verbally, their written accounts differed (see Fig. 4). Participant A did not finish writing a solution when noticing that it did not add up as expected. E wrote an addition that matched the pre-calculated result and D wrote a word answer to the question. Even participant B counted aloud raising a finger on each accentuated word, and got another result: "*thirty-nine, forty, forty-one, forty-two… four*"; the written addition then matched what had been done while counting out loud (see Fig. 5).

Other statements about what is considered a legitimate method when working were: "*you need time to think*" and "*it needs to be quiet*"—the later a social praxis made visible in the classroom that reinforces the idea of difficulty.

Students also expressed a perception that practising skills lead to improvement. One student said: "*I have to confess; I don't know how to solve this; I find it really difficult because I don't know how to do it. So, I don't know how to count it, should I do plus or minus? I don't know, but I'm not going to be ashamed of it; I know I have a little trouble with this, and I should practice*". This perception also highlights self-awareness as a person responsible for one's own knowledge acquisition and feelings about mathematics. Other students' declarations, showing the same self-awareness, are often connected to statements of evaluation, like "*I'm not good at math*", "*I don't like math so it's not fun. If you can [do math], it's more fun*" or "*[y]ou have to practice to become smart*".

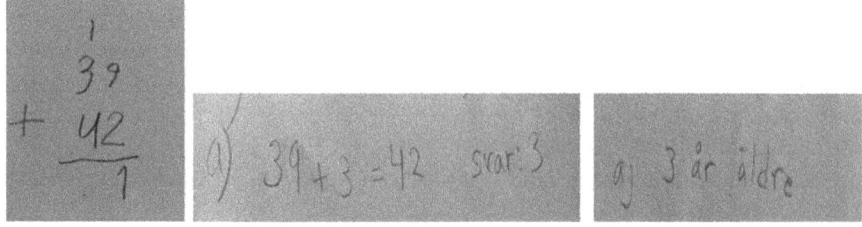

Fig. 4 Solutions to task 164a (see Fig. 2). Reading from left to right: Student A [incomplete], Student E ["answer: 3"] and Student D ["3 years older"]

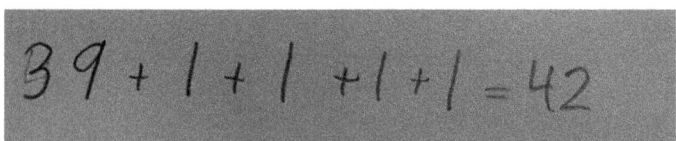

Fig. 5 Student B solution to task 164a (see Fig. 2)

6.4 Cultural Knowledge

Cultural knowledge can be represented in the cultural idea of how you are meant to demonstrate the ability to reason and apply mathematics in real contexts (see Figs. 4 and 5). Or as we could see in Task 165 (see Figs. 2 and 3) even the activity of reading a map can be involved when working with a word problem. It was also shown that mathematical words like distance [*sträcka*] on a map caused trouble because it was read as *streck* [line], a word that has a similar pronunciation and almost the same spelling. Moreover, lines were drawn on the map to show distances (see Figs. 2 and 3) and the word *sträcka* in Swedish also means stretch and reach, so students had to figure out what the word meant.

When students solved Task 736 (see Fig. 6) we discovered something else. Participant B read the word problem out loud, looked at the illustration, said, "*I can't do it*" and did not explain why. Two other participants also reacted to this problem, especially the everyday words in the question. The words *serietidningar* (comic books) and *tidningar* (magazines) appeared in the word problem but the illustration used the word *serier* (series).

Student F said aloud, *Peter bought 16 magazines, where are the magazines? Is it magaz… no those are books… series, where are the magazines? Is it any magazines? Here it says series; is it the same thing? It can't be a series you watch on TV, or…?*

While student A reasoned that "*there [were] different words in the same task!? Can a series be magazines?*", the three different words all referred to the same comic books, thus confusing the students.

In Swedish, *serietidningar*, *tidningar* and *serier* can refer to the same kind of publication. But in many other languages, it would be different. In English, for example, comic books and magazines are not the same kind of publication. Even though the word was not a mathematical term, it took effort and time for the participants to figure out how to solve the task.

One unspoken norm and activity is that the students processed the word problems in the language of instruction. They all interacted with the word problems in Swedish. In the interviews, five of the students said they would not like to have word problems in their first language because they found Swedish easier to read and understand,

735. Christina bought 14 oz of candy. How much did she have to pay? **736.** At one stand comic books are sold. Peter bought 16 magazines. How much did he have to pay?	**735.** Christina köpte 4 hg godis. Hur mycket fick hon betala? **736.** I ett stånd såldes gamla serietidningar. Peter köpte 16 tidningar. Hur mycket fick han betala?

Fig. 6 The market scene, with the word *serier* up to the left (Undvall, et al., 2011), words underlined by us

while three participants replied that it would be nice to have both. Of those three, two clarified that it was not because they did not understand it in Swedish, but for the opportunity to practise their first language. One of them said it would be nice because it would probably be easier to understand than Swedish. The students also said that word problems require that they understand what they read and a knowledge of which calculation methods to use. They claimed that "*you practice both mathematics and Swedish*".

7 Discussion

7.1 The Social

Working with an illustrated word problem is a social experience, and the students' work with the tasks showed how they drew on their former experiences of reading and working with word problems. When they interacted with both the text and the illustrations, they posed more questions to the text but retrieved information from both, showing they were aware of illustrations as a source that could be connected to or combined with the text (Teledahl & Olsson, 2021). But there were some specific situations that interrupted their work, as when a question could not be answered, an illustration was neglected, or a calculation did not add up as intended. Students' experiences related to their (un)familiarity with the procedures of working with mathematical word problems. So, even if the student's work showed they had met word problems earlier, they also showed a gap in their knowledge of the conditions required when working with illustrated word problems. The social activity needed when working with illustrated word problems is a mixture of normative academic practice and life experiences (Barwell, 2018).

The students explained that they wanted to show their thoughts mathematically in writing when working with word problems but were not always sure how to do that. This challenge to communicate mathematical ideas and concepts confirms what Barwell (2009) and Clarkson (2007) found in their studies. Students said that interacting in smaller groups and having the teacher explaining difficult words enabled better understanding. So, they can benefit from collaboration. Students are familiar with posing questions, and they have experienced it as a good way to learn. This means that students could be given more opportunities to talk to each other about the tasks and together clarify words and illustrations in the word problems they are working on.

When not knowing which arithmetic to use for solving a word problem most students tended to say they were not good at math; they took on a specific role, the role of the student who is weak in mathematics. However, some of those students also said they felt responsible for practising their own skills or cultivating their own learning; it could indicate that the student rejects a certain role imposed on those who cannot solve a mathematical problem (Gutiérrez, 2013).

7.2   The Cultural

Students' statements reveal that they perceive a classroom culture where working individually is the established way—the social praxis—of working with word problems in their classroom. We did not do classroom observations; our statements are based on the student's perceptions. And they are expressing a cultural idea and a norm (Barwell, 2018) of how to work with word problems, namely, that it is good to work individually because it makes it easier to think. And, therefore, the need of silence in the classroom seems obvious to them. However, their experience is that they need to speak to make the illustrated word problems understandable, and it becomes noticeable when the words are perceived as difficult, as well as when the context in an illustration is unfamiliar.

It could be that when the students are familiar with the cultural context encapsulated in the illustrated word problems, they have the preconditions needed for working, and then silence becomes natural.

The above findings affirm that word problems always mobilise exclusion. The struggles with solving word problems and unfamiliarity with contexts is what Cooper and Dunne (2000) wrote about years ago. Students' backgrounds play a key role and solving mathematical word problems can sometimes be a test of who the students are. Our participants were not familiar with Kolmården, a typical place for middle class (Swedish) parents to take their children to in summer. We believe the unfamiliarity with the context shows the importance of knowing about students' earlier experiences so the right support can be given to them when solving illustrated word problems. In the case of Kolmården, the context can be explained, and the students could be encouraged to talk about their experiences of zoos.

7.3   The Linguistic

Although everyday words can be troublesome and it takes time to figure out their meaning, most of the students took the time to work through this process. However, this meant that less time was spent on the mathematical work demanded by the word problem, which is consistent with Clarkson's (2009) findings. The students' previous experiences of everyday and mathematical words had an impact on the way they understood and took on the work. The students showed that they were able to draw on experience to work out the meaning of a word, for example, through the process of elimination, working out that, given the context, the word *serier* could not refer to TV series. But drawing on own experiences can be hard when tasks are connected to specific Swedish cultural phenomena (e.g. people, cities, places) that students are not familiar with. The social praxis is interrupted, and a gap appears within the norms. A single word, like a homonym that represents different concepts, then causes confusion, like the Swedish word *sträcka,* which may not be part of students' everyday vocabulary. It can therefore be tricky to know whether students are struggling with the mathematical words, or their second language, or whether they do not understand a concept. As Schleppegrell (2007) states, there are always linguistic challenges when learning mathematics in a second language.

The students also said they found word problems complicated and trickier than mathematical tasks without words. Their statements reveal that it is because they find some words in the problems confusing and it takes time and effort to figure them out. It could mean that the students would benefit from connecting to their first languages. Employing students' mother tongues could adapt teaching to better suit students' needs, to promote their knowledge development based on past experiences and language as recommended by the Swedish curriculum (Skolverket, 2011).

The student said that word problems helped them practice and improve both mathematics and Swedish. And it is not surprising that students find Swedish easier because Swedish has always been their language of instruction and they have always had Swedish textbooks. Furthermore, none of the eight students had attended school elsewhere. The social expectation is that they should be able to solve word problems in Swedish. It is a normative expectation relating to Swedish classroom culture (Norén, 2015) and a sign of their adaptation to the norm of using "only" Swedish in the classroom.

8   Concluding Remarks and Recommendations for Policy

By using the four-fold structure of social praxis as a theoretical framework, we were able to identify the experiences multilingual students mobilised and created when solving illustrated mathematical word problems in Swedish.

Understanding the work with illustrated word problems as social praxis showed that multilingual students experiences related to school culture, having solved word problems before, and remembering what teachers have told them. They tended to rely on previous experiences of solving word problems even when encountering difficulties. The students seemed to skim over illustrations and sometimes even the context because they were so focused on solving the task. They tried to use their mathematical skills. However, neither their experience with Swedish nor their social experiences seemed sufficient. A precondition of working with word problems could also be to address the cultural experiences of multilingual students more thoroughly. If the latter are acknowledged in an everyday school context, they should be part of the learning activities.

We conclude that multilingual students doing mathematics in a language which is not their first language need more opportunities to work with others and communicate verbally, instead of being asked to work individually and quietly. When students ask spontaneous questions to texts, it is possible to take advantage of this. Students must be allowed to mathematise texts and illustrations collaboratively before working individually. Students might also benefit from being allowed to use their first languages (not simply translations) in problem-solving situations, since much of their life experience is related to their daily lives when they are using their first languages (Barwell,

2009; Planas & Civil, 2013). Students need rich opportunities to enhance their experiences of how to solve illustrated word problems. Students themselves refer to the understanding of "how" to solve and "which" arithmetic to use. Active use of language/s is one way to make sure everyone is involved and shares their experience with problems and how to solve them. In this way teachers can adapt their support to the needs of the students, since it becomes easy to capture students' cultural experiences in the classroom's social praxis, and to build new experiences.

The students seemed to exhibit a strong desire to follow normative social praxis relating to "doing" the arithmetic rigorously and to using a method that they believed the teacher would approve of. To increase students' opportunities to experience everyday words in their second language, we suggest combining different school subjects and working thematically. A theme such as Kolmården, for example, could fit particularly well with Geography, where students could elaborate on meaning making, and thus relate words to maps of Sweden and other countries. This theme could also be linked to Orienteering[1] in sports where students get to learn common words, and the experience of Orienteering, in real life. One could also integrate the subject Swedish into Mathematics, using illustrations to elaborate on different words, concepts and homonyms, and examine the word problem genre in relation to other genres which the students encounter in Swedish. The aim would be to explore ways to visualise and make natural connections among mathematical experiences in everyday life by integrating different school subjects.

In line with the Swedish curriculum, there are opportunities for teachers to develop cooperation with mother-tongue-speaking teachers and mentors. Support from these teachers could help students reflect on mathematical issues in more than one language (Swedish). Moreover, access to one's mother tongue can highlight everyday experiences that are created in students' homes, where they speak their mother tongue. Mother tongue also relates to students' cultural backgrounds and to their parents' practices of learning and doing mathematics (Civil et al., 2005).

Our conclusion is that, as multilingual students participate in the social praxis of solving illustrated word problems, they increase their success in relating their cultural experiences to the cultural contexts in the problems. Support for students should include working explicitly with language in different ways, and making the cultural elements embedded in the word problems visible. Students need to submerge themselves in illustrations, words and contexts, and experience word problems in various ways. By dissecting illustrated word problems, talking about them and looking at how the specific word problems are constructed students can better understand how to go about working with them.

Finally, policies that explicitly work with inclusion of multilingual students by supporting language learning—such as the current Swedish educational policy—need to develop a nuanced view of how language and culture are entangled in the particularities of school practices such as the work with illustrated word problems in mathematics. Explicit inclusive policy formulations are not enough. This may

---

[1]In Sweden a common "competitive sport in which runners have to find their way across rough country with the aid of a map and compass" (https://www.lexico.com/definition/orienteering).

mean that policy could also promote concrete spaces for teachers' *policy enactment*[2] where language, culture and experiences can be discussed and negotiated in the classroom. In the concrete case of illustrated word problems in mathematics, such spaces of policy enactment may mean that it is not sufficient with allowing children to strengthen their mother tongue to understand the problem with the hope that they can produce a mathematically adequate answer. The issue is rather to open the opportunity of recognising the particular cultural norms and life forms that are embodied in the problems—and the curriculum as a whole—and to bring them in contrast and discussion with respect to other possible life experiences emanating from the diversities of students' lives. These spaces of enactment are indispensable, since it is only in the details of praxis that possibilities for real inclusion can be worked out by teachers in classrooms.

References

Ball, S. J., Maguire, M., & Braun, A. (2012). *How school do policy: Policy enactments in secondary schools*. London: Routledge.

Barwell, R. (2009). Mathematical word problems and bilingual learners in England. In R. Barwell (Ed.), *Multilingualism in mathematics classrooms: Global perspectives* (pp. 63–77). Bristol: Multilingual Matters.

Barwell, R. (2018). Word problems as social texts. Numeracy as social practice: Global and local perspectives, (pp. 101–120).

Barwell, R., Wessel, L., & Parra, A. (2019). Language diversity and mathematics education: New developments. *Research in Mathematics Education, 21*(2), 113–118. https://doi.org/10.1080/147 94802.2019.1638824.

Boonen, A. J. H., de Koning, B. B., Jolles, J., & van Der Schoot, M. (2016). Word problem solving in contemporary math education: A plea for reading comprehension skills training. *Frontiers in Psychology, 7,* 191. https://doi.org/10.3389/fpsyg.2016.00191.

Civil, M., Planas, N., & Quintos, B. (2005). Immigrant parents' perspectives on their children's mathematics education. *ZDM Mathematics Education, 37*(2), 81–89.

Clarkson, P. C. (2007). Australian Vietnamese students learning mathematics: High ability bilinguals and their use of their languages. *Educational Studies in Mathematics, 64*(2), 191–215.

Clarkson, P. C. (2009). Mathematics teaching in Australian multilingual classrooms: Developing an approach to the use of classroom languages. In R. Barwell (Ed.), *Multilingualism in mathematics classrooms: Global perspectives* (pp. 145–160). Bristol: Multilingual Matters.

Cooper, B., & Dunne, M. (2000). *Assessing children's mathematical knowledge: Social class, sex, and problem-solving*. Buckingham: Open University Press.

De Corte, E., Verschaffel, L., & Greer, B. (2000). Connecting mathematics problem solving to the real world. In *Proceedings of the International Conference on Mathematics Education into the 21st Century: Mathematics for living* (pp. 66–73).

Dewolf, T., Van Dooren, W., Hermens, F., & Verschaffel, L. (2015). Do students attend to representational illustrations of non-standard mathematical word problems, and if so, how helpful are they? *Instructional Science, 43*(1), 147–171.

---

[2]The notion of policy enactment (Ball et al., 2012) refers to the idea that the most important aspect of a policy is not just the statement of intentions in official documents, but the constant work of translation and recontextualizations of those intentions that teachers do in their everyday work in schools and classrooms.

Doğan, O., & Haser, Ç. (2014). Neoliberal and nationalist discourses in Turkish elementary mathematics education. *ZDM Mathematics Education, 46*(7), 1013–1023.

Dyrvold, A., Bergqvist, E., & Österholm, M. (2015). Uncommon vocabulary in mathematical tasks in relation to demand of reading ability and solution frequency. *Nordic Studies in Mathematics Education, 20*(1), 101–128.

Elia, I., & Philippou, G. (2004). The functions of pictures in problem solving. In *International group for the psychology of mathematics education*. International Group for the Psychology of Mathematics Education, 35 Aandwind Street, Kirstenhof, Cape Town, 7945, South Africa.

Fan, L., Xiong, B., Zhao, D., & Niu, W. (2018). How is cultural influence manifested in the formation of mathematics textbooks? A comparative case study of resource book series between Shanghai and England. *ZDM Mathematics Education, 50*(5), 787–799.

Gerofsky, S. (1996). A linguistic and narrative view of word problems in mathematics education. *For the Learning of Mathematics, 16,* 36–45.

Gerofsky, S. (2004). *A man left Albuquerque heading east: Word problems as genre in mathematics education* (Vol. 5). Peter Lang.

Greer, B. (1993). The modeling perspective on wor(l)d problems. *Journal of Mathematical Behavior, 12,* 239–250.

Greer, B. (1997). Modelling reality in mathematics classrooms: The case of word problems. *Learning and Instruction, 7*(4), 293–307.

Gutiérrez, R. (2013). The sociopolitical turn in mathematics education. *Journal for Research in Mathematics Education, 44*(1), 37–68.

Lave, J. (1992). Word problems: a microcosm of theories of learning. In P. Light, & G. Butterworth (Eds.), *Context and cognition: Ways of learning and knowing*. Hemel Hempstead: Harvester Wheatsheaf.

Moschkovich, J. (2007). Using two languages when learning mathematics. *Educational Studies in Mathematics, 64*(2), 121–144.

Norén, E. (2010). *Flerspråkiga matematikklassrum: Diskurser i grundskolans matematikundervisning*. Doctoral dissertation, Stockholms University.

Norén, E. (2015). Agency and positioning in a multilingual mathematics classroom. *Educational Studies in Mathematics, 89*(2), 167–184.

Norén, E., & Caligari, L. (2021). Practices in multilingual mathematics classrooms: Word problems. In L. Björklund Boistrup, J. Häggström, Y. Liljeqvist, & O. Olande (Eds.), *Sustainable mathematics education in a digitalized world. MADIF 12, The twelfth Research Seminar in Mathematics Education by SMDF*. Växjö: Linnéuniversitetet.

Planas, N., & Civil, M. (2013). Language-as-resource and language-as-political: Tensions in the bilingual mathematics classroom. *Mathematics Education Research Journal, 25*(3), 361–378.

Prediger, S., & Schueler-Meyer, A. (2017). Fostering the mathematics learning of language learners: Introduction to trends and issues in research and professional development. *EURASIA Journal of Mathematics, Science and Technology Education, 13*(7b), 4049–4056.

Radford, L. (2008a). Culture and cognition: Towards an anthropology of mathematical thinking. In *Handbook of international research in mathematics education* (2nd ed., pp. 439–464).

Radford, L. (2008b). The ethics of being and knowing: Towards a cultural theory of learning. In L. Radford, G. Schubring, & F. Seeger (Eds.), *Semiotics in mathematics education: Epistemology, history, classroom, and culture* (pp. 215–234). Rotterdam: Sense Publishers.

Radford, L. (2018). Semiosis and subjectification: The classroom constitution of mathematical subjects. In I. N. Presmeg, L. Radford, W.-M. Roth, & G. Kadunz (Red.), *Signs of signification: Semiotics in mathematics education research* (s. 21–35). Springer International Publishing.

Radford, L., & Empey, H. (2007). Culture, knowledge and the self: Mathematics and the formation of new social sensibilities in the Renaissance and medieval Islam. *Revista Brasileira de História Da Matemática. An International Journal on the History of Mathematics, 1,* 231.

Ryan, U. (2019). *Mathematics classroom talk in a migrating world*. Doctoral dissertation, Malmö University.

Schleppegrell, M. J. (2007). The linguistic challenges of mathematics teaching and learning: A research review. *Reading & Writing Quarterly, 23,* 139–159.

Skolinspektionen (2010). Språk- och kunskapsutveckling för barn och elever med annat modersmål än svenska. (Kvalitetsgranskning rapport 2010:16). Stockholm: Skolinspektionen. Från http://www.skolinspektionen.se/Documents/publikationssok/granskningsrapporter/kvalitetsgranskningar/2010/sprakutveckling-annat-modersmal/kvalgr-sprakutvslutrapport.pdf.

Skolverket. (2011). *Läroplan för grundskolan, förskoleklassen och fritidshemmet 2011.* Stockholm: Skolverket.

Skolverket. (2017). *Kommentarmaterial till kursplanen i matematik.* Stockholm: Skolverket.

Skolverket. (2019a). *"Betyg i grundskolan årskurs sex vårterminen 2019".* https://www.skolverket.se/skolutveckling/statistik/sok-statistik-om-forskola-skola-och-vuxenutbildning.

Skolverket. (2019b). *PISA 2018. 15-åringars kunskaper i läsförståelse, matematik och naturvetenskap.* Stockholm: Skolverket.

Souza, D., & da Silva, M. A. (2018). O dispositivo pedagógico do currículo-brinquedo de matemática, marcado pela dimensão de gênero, na produção de subjetividades. *Reflexão e Ação, 26*(2), 149–164.

Svensson Källberg, P. (2018). *Immigrant students' opportunities to learn mathematics: In(ex)clusion in mathematics education* (PhD dissertation). Department of Mathematics and Science Education, Stockholm University, Stockholm.

Teledahl, A., & Olsson, J. (2021). Students' use of written and illustrative information in mathematical problem solving. In L. Björklund Boistrup, J. Häggström, Y. Liljeqvist, & O. Olande (Eds.), *Sustainable mathematics education in a digitalized world. MADIF 12, The twelfth Research Seminar in Mathematics Education by SMDF.* Växjö: Linnéuniversitetet.

Undvall, L., Melin, C., Ollén, J., & Welén, C. (2011). *Alfa åk 4.* Stockholm: Liber.

Laura Caligari is a Ph.D. student in the Department of Mathematics and Science Education at Stockholm University. Her doctoral research focuses specifically on mathematical word problems and multilingual learners. Laura has worked as a teacher in elementary school for over a decade, mainly with multilingual learners in Sweden.

Eva Norén is Associate Professor at the Department of Mathematics and Science Education, Stockholm University. Her research focuses on multilingual students in mathematics classroom, as well as social, critical and political issues in mathematics education.

Paola Valero is Professor of Mathematics Education at the Department of Mathematics and Science Education, Faculty of Science, Stockholm University, Sweden. Her research interests are mathematics education at all levels; in particular policy enactment processes, curricular development, multiculturalism and multilingualism in mathematics education, and diversity in mathematics teacher education. Currently, her research explores the significance of mathematics and science education as fields where power relations are actualized in producing subjectivities and generating in(ex)clusion of different types of students. She is currently the leader of the Swedish Research Council funded national Graduate school "Relevancing mathematics and Science Education (RelMaS)". Her research has been part of the Nordic Center of Excellence "Justice through education in the Nordic Countries".

Language Policy for Equity in University STEM Education in Postcolonial Contexts: Conceptual Tools for Policy Analysis and Development

Kate le Roux, Bongi Bangeni, and Carolyn McKinney

Abstract This chapter is concerned with university language policy for equity in STEM education in postcolonial contexts with diverse language landscapes. This focus is necessary, given how language acts to enable or challenge inequity in particular historical and geo-political relations, and the need for research on university language policy, specifically for STEM education, to complement research on practice. We propose conceptual tools for analysing and developing policy. We view language and STEM as historical, social and political practices and, following Hilary Janks and Rochelle Gutiérrez, equity as having two dominant meanings (*access, achievement*), and three critical meanings (*power, diversity, design*). We illustrate the potential use of these tools in a critical discourse analysis of the language policy of a South African university. This analysis shows a policy focus on access to and achievement in dominant STEM knowledge in 'English', with some attention to diversity and power in representations of language for STEM and the language-user. We end with five recommendations for future policy development. We position this chapter as an example of language policy analysis that responds to the specificity of context, but which potentially makes a theoretical contribution beyond the context in a way that does not universalise.

Keywords Equity · Higher education · Language policy · Multilingualism · Postcolonial

K. le Roux (✉) · B. Bangeni · C. McKinney
University of Cape Town, Cape Town, South Africa
e-mail: kate.leroux@uct.ac.za

B. Bangeni
e-mail: abongwe.bangeni@uct.ac.za

C. McKinney
e-mail: carolyn.mckinney@uct.ac.za

© The Author(s), under exclusive license to Springer Nature Switzerland AG 2021   195
A. A. Essien and A. Msimanga (eds.), *Multilingual Education
Yearbook 2021*, Multilingual Education Yearbook,
https://doi.org/10.1007/978-3-030-72009-4_11

1   Introduction

This chapter is concerned with university language policy in postcolonial contexts with diverse language landscapes. Our specific focus is how language policy—as historical, social and political text, with material effects—may enable or constrain equity in STEM education. We use 'STEM' to include disciplines in Science, Technology, Engineering, Mathematics and Health Sciences. We propose tools for conceptualising language, STEM knowledge and equity, and demonstrate their use in a critical discourse analysis of the language policy of the University of Cape Town (UCT), an elite, historically 'white', 'English'-medium, public South African university.

In this introduction, we use the literature to motivate for our focus, and further develop this motivation in subsequent sections. Firstly, since notions of 'language', 'multilingualism' and 'science' are constructed in particular historical and geopolitical relations (García & Lin, 2018; Makoni & Pennycook, 2007; Prah, 2017), there is a need to attend to the context-specificity of language policy for STEM. We write *from a postcolonial context*, recognising that such contexts are themselves diverse and changing. Historically, 'language' and Euromodern 'science' were part of the "cultural kitbag" (Bishop, 1990, p. 58) of colonial practices—and education in particular—that constructed inequities, for example, through their use to distinguish different groups of people according to their 'worth' on a hierarchy (Bishop, 1990; Glissant, 2010). We illustrate this point in our description of the South African context in Sect. 4. Given the enduring dominance of colonial languages, both locally and globally, postcolonial universities seeking to avoid (re)producing language inequities need to (a) interrogate how history shapes current language policy and practice, and with this understanding (b) fundamentally rethink inherited conceptions and experiences of language to transform physical and knowledge spaces (Cele, 2004; Makoni & Pennycook, 2007). Exploring language for equity in university STEM in such contexts requires a concept of equity that is more sophisticated than considerations of access to and success in the dominant knowledge using historically dominant languages.

Secondly, there is a need for research specifically on language *policy*, to complement research on *practice*. Cases of 'multilingual' practice in university courses in South Africa, including in STEM (e.g. Dalvit, 2010; Leeuw, 2014; Madiba, 2019) have been reported. Yet research (e.g. Cele, 2004; Kotzé, 2014) and policy (Department of Higher Education and Training [DHET], 2020) identify a lack of progress in reshaping the language landscape of South African universities. This literature proposes political, economic, social, linguistic and managerial reasons for poor policy implementation. While these arguments cannot be disputed, there is growing recognition that language policy itself needs to be problematised, evidenced by detailed case studies of institution-specific language policy (e.g. Nudelman, 2015; van der Merwe, 2016), as well as in South African students' calls for 'free decolonised education' voiced in protest action since 2015 (e.g. Gillespie & Naidoo, 2019). These protests and recent scholarship (e.g. Luckett, 2016) suggest that efforts to transform South African universities need to look beyond institutional structures, to the wider Euromodern and neoliberal ideologies that underpin institutions such as UCT. Language ideologies, specifically, have been the focus of recent scholarship on language policy and practice in South African schooling (e.g. Makoe & McKinney, 2014; McKinney, 2017), and we extend that work to *university language policy*.

Thirdly, there is a need to focus on language policy for university *STEM* education. Drawing theoretically from the sociology of education and social semiotics, scholars have identified differences in the nature of knowledge and language use across broad fields of 'Science', 'Social Science' and 'Humanities', as well as between disciplines (e.g. Dalvit, 2010; Kuteeva & Airey, 2014). Writing from a European context, Kuteeva and Airey (2014) use these differences to argue for disciplinary-specific language policies. Yet there is a need to explore what such a policy might look like in postcolonial contexts. In South Africa, policy development can be informed by a growing body of work on the intellectualisation of named African languages in quantitative disciplines such as computer science, economics, mathematics, psychology and statistics, and its use in university classrooms (e.g. Dalvit, 2010; Madiba, 2014, 2019; Mkhize et al., 2014; Paxton, 2009; Whitelaw et al., 2019).

To respond to the three needs identified here, we propose—as described in detail in Sect. 2—a particular conceptualisation of language and STEM knowledge as intricately related to power in social practices. We view language policy as historical, social and political text, related to other texts such as institutional strategic plans and national university language policy. Drawing on Hilary Janks (2010) and Rochelle Gutiérrez (2002, 2012), we adopt a five-part notion of equity, comprising *access*, *achievement*, *power*, *diversity* and *design*. These conceptual tools can be used to ask how language policy in a postcolonial context may enable or constrain equity in university STEM education. Specifically, this involves answering sub-questions about how policy text:

- represents 'language', 'multilingualism', and 'STEM knowledge', and their relations;
- identifies the purpose and location of language use;
- identifies the language-user;
- constitutes and locates the 'language problem' to which policy responds;
- constitutes the solution to the problem and
- asserts its status at the institution.

In this chapter, we use the case of language policy at UCT to illustrate the use of our conceptual tools to answer these questions. The chapter is structured as follows: We describe and motivate for our conceptual tools (Sect. 2), describe our methodological approach to the case (Sect. 3), locate the case in the context of postcolonial South Africa (Sect. 4), apply our tools to analyse UCT policy (Sect. 5) and conclude by demonstrating how an analysis such as this can inform language policy development (Sect. 6).

2 Conceptual Tools

2.1 *Conceptualising Language and STEM Knowledge*

Policy falls between ideology and practice (Shohamy, 2006, cited by van der Merwe, 2016). Thus, to understand what is identified as the 'language problem' to which policy responds, the ideologies of language and STEM knowledge that underpin policy need to be identified. By ideology we mean "the sets of beliefs, values and cultural frames that continually circulate in society, informing the ways in which language [and knowledge] is conceptualized as well as how it is used" (Makoe & McKinney, 2014, p. 659). We distinguish two broad conceptualisations of language and STEM knowledge in the literature.

Firstly, STEM knowledge may be viewed as objective, neutral, bounded and fixed, and thus universal and transferrable unproblematically across contexts. Mathematics is an excellent example making it an ideal base for scientific knowledge; in mathematics 'truth', what counts as mathematical knowledge, is intrinsic to the logic of the discipline and judged by the rigour of proof, and thus it can represent the essence of all things in an objective manner (Bishop, 1990; Skovsmose, 2016; Gutiérrez & Dixon-Román, 2011). Similarly, if language is regarded as neutral, unitary, bounded and stable, it can be viewed as an object that can be standardised in lexical and grammatical rules. Thus, language can 'carry' fixed STEM meanings across contexts. This *monoglossic* ideology normalises the naming of languages such as 'English' or 'isiXhosa', and as 'first'/ 'second' languages, the practice of 'code-switching', and 'multilingualism' as the adding of named languages (Makoni & Pennycook, 2007; McKinney, 2017). García and Lin (2018) refer to the last-mentioned as *elite multilingualism*, as distinct named languages are hierarchised. A monoglossic ideology informs the naming of language as 'scientific'/ 'everyday' (Tyler, 2016), as 'developed'/ 'undeveloped' for science, and for those languages that are 'developed', the view that scientific ideas can be unproblematically translated across these languages. This perspective of language normalises monolingualism, and locates the 'language problem' in the 'multilingual' student who is not 'proficient' in the 'standard', dominant, named language for STEM.

A second approach views STEM disciplines as historical, social and political practices involving identifiable combinations of knowledge, activities, technologies, social relations, values, identities and language use (Fairclough, 2003). Here, language has an ontological, epistemological and relational function in STEM. So, the focus of this perspective is language use by people in practice (in 'translanguaging'), language as changing and developing in use (Finlayson & Madiba, 2002), and language as *heteroglossic*, that is, "the complex, simultaneous use of a diverse range of registers, voices, named languages or codes" which form part of a multimodal repertoire for meaning-making in a particular context (McKinney, 2017, p. 22, following Bakhtin, 1981). This ideology normalises *indigenous multilingualism*, that is, how the majority of people in postcolonial contexts grow up using and continuing to use various named languages flexibly (García & Lin, 2018). Here, the 'multi-lingual' student uses a rich repertoire of language practices in different roles and contexts. Language policy is historical, social and political text, that works ideo-logically to produce what is 'normal' language use at university, and has material effects.

From this perspective, the named languages and 'science' of postcolonial contexts are 'invented' in coloniality, that is in dialectical, asymmetrical interaction between colonisers and the colonised. Historically, European languages were drawn on to codify indigenous language use in writing as named, bounded languages (García & Lin, 2018; Makoni & Pennycook, 2007), and a binary has been produced between '(Euromodern) science' and 'indigenous knowledge'. 'Euromodern science' in colo-nial languages was constructed as authoritative and learned, and used as a tool for governmentality.

2.2 Conceptualising Equity

Equity is often used interchangeably with 'fairness', 'democratic access' and 'justice', as distinct from 'inequality' as 'sameness' of opportunities or outcomes (Gutiérrez, 2002; Pais, 2012), and also as in tension with 'excellence' and 'quality'. We propose a five-part concept of equity that draws from scholarship in contexts of language diversity: Janks' critical literacy and Gutiérrez's sociopolitical perspective of mathematics education. The five meanings are not new. Yet we demonstrate in this chapter that, taken together, they respond to a recognised need for a nuanced concept of equity that brings into view, in a particular context, not just certain groups or individuals but the system itself and the underpinning ideologies (Gutiérrez, 2002; Pais, 2012).

Our five-part concept of equity has two dominant meanings: language use for *access* to, and *achievement* in the dominant STEM knowledge. Gutiérrez (2012) suggests these are about "playing the game" (p. 21) in the current status quo. *Access* is commonly viewed as "opportunities to learn" in the form of "tangible resources" such as good teachers, and quality curriculum and learning materials (p. 19). Yet viewing equity in education as a didactical issue has not yielded much gain (Pais, 2012). In a study of mathematics education for Health Sciences in South Africa, le Roux and Rughubar-Reddy (forthcoming) argue for a broader notion of access. Firstly, *formal access* to academic programmes by meeting language and STEM entrance requirements, to financial support for tuition and living costs, and to safe accommodation and productive learning spaces. Secondly, *epistemic access* (or *epistemological access* for Morrow, 2009) to the valued STEM knowledge in the domi-nant language of teaching, learning and assessment (LoLT). This includes being listened to by influential audiences (Janks, 2010, citing Bourdieu, 1991), and being heard when using language to ask questions and to demonstrate one's learning. Lastly, *social access* is "the possibility to inhabit a space to an extent that one can say, 'This is my home. I am not a foreigner. I belong here'" (Mbembe, 2016, p. 30).

The second dominant meaning of equity, *achievement*, is about student success in the dominant STEM knowledge in the LoLT, as measured in course taking patterns, assessments, accreditations and participation in the "pipeline" (Gutiérrez, 2012, p. 19).

The three critical meanings of equity are about challenging a language and knowledge status quo that (re)produces asymmetrical power relations, and reshaping or "changing the game" (Gutiérrez, 2012, p. 21). Firstly, *power* (or *domination* for Janks) is about recognising language and STEM knowledge as historical, social and political practices and, in postcolonial contexts, 'disinventing' dominant conceptions to understand their historical constitution and (re)production in contemporary times (Makoni & Pennycook, 2007; Mbembe, 2016). The second critical meaning of equity is *diversity* (or *identity* for Gutiérrez) which acknowledges different ways of using language, and related STEM knowledge and identities. But not all notions of 'diversity' are critical, for if knowledge, language, identity are seen as fixed, enumerable objects, differences may either be hierarchised or used in a form of romanticised plurality that reinforces domination (Janks, 2010; Makoni & Pennycook, 2007). However, if language, knowledge and identities are seen as practised and hence changing—including a view of indigenous languages as growing as languages of science (Finlayson & Madiba, 2002)—then diversity is a productive resource for change.

Yet it is not enough to disinvent dominant views of language and STEM knowledge, or to identify in diversity better descriptions thereof, hence the third critical meaning, *design*. This involves destabilising what is 'normal', expanding what counts as legitimate language use for STEM, and recognising new meanings as necessary in a contemporary, postcolonial world (Janks, 2010; Makoni & Pennycook, 2007).

Crucially, all five meanings are interdependent and equally important for equity. For "changing the game requires being able to play it well enough to be taken seriously" (Gutiérrez, 2012, p. 21). This tension leads to an unavoidable *access paradox* (Janks, 2010) in postcolonial contexts. For example, if we provide access for all to the dominant STEM knowledge in a colonial language, this contributes to maintaining the dominance of these forms, and the potential for design presented by diversity is not realised. Yet if we do not support students to access the dominant forms, we perpetuate historical asymmetries in a society that continues to recognise only these forms.

3   Methodology

3.1   A Case Study of Language Policy

The language landscape of the 26 public South African universities, South Africa and other postcolonial contexts is diverse and changing. Yet, our choice of a case study of one university—in particular UCT which is an elite, 'English'-medium institution with a strong and enduring colonial legacy, as described in Sect. 4—has value in two respects. Duminy et al. (2014) argue that the case study in a postcolonial context offers "nuanced" (p. 10) knowledge for local practice and policy, but can also "bring back" (p. 3) traditionally periphery contexts for wider theoretical contributions. Thus, our intention in this chapter is to offer a contribution that neither essentialises the local case, nor claims universality.

3.2   A Critical Discourse Analysis of UCT Language Policy

UCT developed its first language policy in 1999, with revisions in 2003 and 2013. A new policy is in progress. Thus, in this chapter we analyse the most recently published 2013 Policy (UCT, 2013a) and the related Draft Implementation Plan (UCT, 2013b),[1] acknowledging their production at that time in UCT's history. In the absence of a more recent published policy, we search in more recent institutional and faculty planning texts for signs of current thinking about language that might inform an upcoming policy.

We analyse these policy documents using Fairclough's (2003) method of critical discourse analysis, which is aligned with our conceptualisation of language as described in Sect. 2. From this perspective language policy text is dialectically related to the wider historical, social and political practices in which it is located. On the one hand, the text gives meaning to or constitutes 'language', 'knowledge', the 'language-user' and so on. On the other hand, policy text is itself shaped by "circumstances, histories, trajectories, strategic positions and struggles within these contexts" (Fairclough, 2006, p. 167).

Fairclough's (2003) method of critical discourse analysis involves working to-and-fro between three levels of analysis. At the micro-, sentence-level of *description,* we perform a content and linguistic analysis of the texts. We analyse the lexical and grammatical choices (such as nouns and adjectives for naming languages and language-users, verbs and modality for the policy tone), the order and extent of coverage of ideas, and the warrants for claims to legitimacy. At the meso-level of *interpretation,* we look across the texts to ask, What meanings are present/absent? What meanings are foregrounded/backgrounded? At the macro-level of *explanation,* we consider how the texts might be shaped by the wider context; we look for traces of the conceptualisations of 'language', 'knowledge' and so on available in the wider context, and described in Sects. 2 and 4.

---

[1]The Draft Implementation Plan (UCT, 2013b), developed by the Language Policy Sub-committee (LPC), was provisionally approved by the University Senate Teaching and Learning Committee, subject to costing. This Plan was made available for this analysis by LPC member Carolyn McKinney.

4   The Context of the Case of UCT Language Policy

4.1   *Language and STEM in Education*

Formal colonial rule by the Dutch and then the British from 1652 to 1948 set in place particular racial, social, economic, knowledge and linguistic hierarchies in South Africa. Education for the colonisers, not the indigenous peoples, was prioritised, with mission schools educating a small black African elite in 'English'. UCT, founded in 1829, has since 1928 been located on land bequeathed by the British imperialist Cecil John Rhodes. The language 'Afrikaans' was developed from Dutch, with Portuguese, Indonesian, Malay and local Khoisan influences. Colonisers codified named African languages such as 'isiXhosa' in written form in genres such as religious but not scientific texts.

During apartheid (1948–1994) colonial hierarchies were entrenched legally, spatially and institutionally. Education for students legally classified as 'white' was provided in a student's 'home' language of either 'English' or 'Afrikaans' (with the development of Afrikaans as an academic and scientific language prioritised), and focused on knowledge, including STEM, for academic and skilled labour. For those classified as 'black African', policy dictated named 'African' languages as the LoLT in primary schools, with a switch to 50–50 'English'-'Afrikaans' in high school. STEM knowledge was not regarded as necessary for those being schooled for unskilled labour.

The interaction between this sociopolitical history, modernity and neoliberalism presented a newly democratic South Africa with multiple challenges such as redressing past injustices, developing an inclusive, democratic nation and meeting local social and economic needs in a competitive neoliberal, globalised world. The 1996 constitution recognises 11 named, 'official' languages: 'Afrikaans', 'English', 'isiNdebele', 'isiXhosa', 'isiZulu', 'Sepedi', 'Sesotho', 'Setswana', 'siSwati', 'Tshivenda' and 'Xitsonga'.

Twenty-five years into democracy, the UCT student population is more diverse racially and linguistically, as measured by self-declared apartheid racial classification and 'home' language. Yet UCT continues to grapple with its strong and enduring colonial legacy, recognised in its dominant Euromodern, 'white' and 'English' institutional and knowledge structures. Student protests since 2015 highlight how historically marginalised students feel they do not belong in such spaces (Gillespie & Naidoo, 2019). Research on STEM achievement at UCT shows that proficiency in 'English' matters in complex interplay with race, class, geography and schooling (e.g. Rooney, 2015).

The pipeline to achievement in university STEM needs to be understood in the context of schooling; the majority of school students study STEM in a language they are not proficient in. Less than one-sixth of South Africans report using 'English' inside or outside of the household (Statistics South Africa, 2019). Yet the school curriculum promotes an early exit model of 'bilingualism' from 'home' language as

LoLT in Grades 1 to 3 to 'English' or 'Afrikaans' as LoLT in Grade 4 (McKinney, 2017), and the subject 'English Additional Language' is cognitively undemanding (Kapp & Arend, 2011).

4.2 Language Policy in South African Universities

South African Universities are required to develop their own language policies and implementation plans, in line with the constitution and national policy. The first national Language Policy for Higher Education (LPHE) was gazetted in 2002 (Ministry of Education, 2002). A revision was gazetted in 2020 (DHET, 2020), for implementation in 2022.

The two policies are similar in the following respects: "equity" is used alongside "equality" and "fairness" (DHET, 2020, p. 13) to refer to "official", named languages being used at "multilingual" universities; both recognise the political nature of language in South Africa, language as a "barrier" to university access and success, and the need for resources to develop indigenous languages; and both implicitly suggest a tension between working within and challenging the status quo.

Yet the two policies differ, firstly, in that the 2002 LPHE offers more space—in the short term—for working within the "status quo" (Ministry of Education, 2002, p.10) to support students to learn in the colonial languages as LoLTs, while also "promoting" (p. 14) 'multilingualism' for institutional transformation. The development and use of named indigenous languages—a medium-term to long-term goal (pp. 3–4)—for "equity" and "redress" is balanced against "practicability" and individual constitutional "rights". In contrast, the revision represents indigenous languages as "meaningful academic discourse" (DHET, 2020, p. 9), and important for "cognitive and intellectual development" (p. 5).

Secondly, the 2002 LPHE makes one reference to "academic literacy" (Ministry of Education, 2002, p. 11) in the LoLT and does not recognise disciplinary-specific literacies. The revision defines "academic language" as having "discipline-specific vocabulary, grammar, punctuation, argumentation and discourse" and "rhetorical conventions" (DHET, 2020, p. 7) and stresses the potential of indigenous languages to function as "sources of knowledge in the different disciplines of higher education" (p. 9). Lastly, there is a shift in where the 'language problem' is located and hence the tone of each policy. The 2002 LPHE uses a language of "encouragement" (Ministry of Education, 2002, p. 15) in "promoting" (p. 14) multilingualism and developing indigenous languages. Yet the 2020 policy shows a level of frustration with universities for not giving indigenous languages "the official space to function as academic and scientific language" (DHET, 2020, p. 9). Thus, it prescribes what universities "must" do, and the need for government monitoring and evaluation.

5 Critical Discourse Analysis of UCT Language Policy for Equity in STEM

5.1 UCT Language Policy (2013)

The substance of the 1.5 page Policy (UCT, 2013a) is on page one. Three, numbered objectives are identified: (1) the development of "multilingual proficiency and awareness"; (2) the "development" of all South African languages for use in instruction, and "promotion" of scholarship in these languages and (3) ensuring that "students acquire effective literacy in English", defined as "the ability to communicate through the spoken and written word in a variety of contexts: academic, social, and professional". Objective one, as the "starting point", locates the 'language problem' in the student who "needs" to "acquire" such proficiency and awareness, with "multilingualism" having "personal, social and educational value". Later the problem is also located in the university's internal and external communication.

The naming of 'English' as the "primary" LoLT suggests the possibility of other LoLTs, while still foregrounding 'English'. Overall, the Policy establishes rather than problematises the *power* of "literacy" in 'English', that is, an ideology of language as monoglossic and of Anglonormativity. 'English' is named with certainty ("English *is*…", our italics) as "the primary language" of teaching and examination "at all levels" (except in language and literature departments) and of governance and administration, and is an "international language". Thus, "educational value" lies in 'English', with *achievement* in 'English' necessary for degree purposes and for participation in a global pipeline. Students need to "acquire effective literacy in English" for *epistemic access*, and university communication in 'English' should be "clear" and "concise" to enable *physical access* for all. Although students need "the spoken and written word" in English in "academic, social, and professional" contexts, reference to the *diversity* of language use in disciplines and the visual and symbolic modes for STEM knowledge is absent. Consistent with national LPHE, UCT Policy is less prescriptive on objectives one and two, than on objective three; academic staff are "expected" to "explore" ways to achieve the former; objective two is a "medium- to long-term" goal; and all three "official" languages of the region—'English', 'isiXhosa' and 'Afrikaans'—are to be "promoted" for communication and used "where practical".

Unlike national LPHE, the Policy does not represent ongoing support for literacies in 'English' as in tension with the development and use of other named languages. The word "multilingual" is used many times, but predominantly as an adjective for "proficiency" and "awareness", and the "social" and "personal", and not "educational", value thereof is developed, that is, for "participation in society". This is an elite multilingualism that should run in parallel to 'English' at UCT. The presence of many staff and students for whom 'English' is not the "primary language" is noted, but this is represented more as a problem than a resource. Language *diversity* is used uncritically to refer to the presence of multiple named languages in the university and society, rather than as a heteroglossic practice for disciplinary meaning-making. Thus, the need and potential for *design* in the sense of challenging the status quo is not surfaced in the Policy.

5.2 UCT Draft Language Policy Implementation Plan (2013)

Our analysis of the 19-page draft Plan (UCT, 2013b), developed by a Language Policy Sub-committee, suggests that it fulfils its mandate to provide strategies, time-lines, responsibilities and funding requirements for the 2013 Policy implementation. Importantly, the Plan makes discursive moves that shift this Policy conceptually.

The Plan (UCT, 2013b) renames and re-orders the Policy objectives: (1) "Academic Literacy in English Strategy" (the original "Literacy" renamed), (2) "Multilingualism Strategy" and (3) "Promote Scholarship in African languages (isiXhosa)" ('Afrikaans' not named). These are linked to the constitution, the 2002 LPHE emphasis on "multilingualism" for "equity of *access* and *success* for all students" (our italics), and to UCT's commitment to *diversity* in the sense of "social justice and democratic values" (p. 1). The first two objectives are identified as "main objectives" which are "essential graduate attributes" (p. 1) for student *achievement*. Initially, the two objectives are represented in tension ("on the one hand […] and on the other […]", p. 1), but they are subsequently named as "intertwined and complement[ary]". Their relations are given meaning in the notion of a "continuum" of a student's *"language and literacy repertoire"* (p. 1, italics in the original). This is defined as "the range of languages and varieties that a person uses to perform particular roles and tasks", with an example for one student provided (Table 1).

The notion of "repertoire" extends the *diversity* of language, challenging the dominant narrative of named languages; language is contextual (used in "clinical" settings and "scientific reports"), includes reading, and languages are not "separate distinct linguistic codes" (p. 2).

The Plan focuses first on a student's "Academic Literacy in English". Drawing legitimacy from international and local scholarship, language is represented as "central" for university learning (p. 2). Indeed, it is here that the initial textual reference to "equity" is developed, with language "cut[ting] to the heart of UCT's equity goals" (p. 4). Again, language *diversity* is expanded: the focus is on "a new variety of language", that "embodies" disciplinary knowledge, values and forms of expression,

Table 1 Representation of a student's "language and literacy repertoire" (UCT, 2013b, p. 2)

Formal curriculum			*Informal interaction*
Academic literacy in English	Multilingualism for learning	Multilingualism for professions	Multilingualism for interaction
e.g. Scientific report, essay, MCQ, thesis	e.g. glossaries	e.g. clinical case histories	e.g. isiXhosa conversation classes

and includes "digital literacies and numeracy" (p. 2). It notes that given the historical and current *power* asymmetries, the use of English as LoLT at UCT "shapes an individual's chances of success" (p. 2). Thus "Academic Literacy in English" is recognised as necessary for *access*.

Describing the 'language problem' and its solution in the section "Academic Literacy in English", the Plan expands the discourse on who needs language support and further expands on *diversity*, while still noting the Anglonormativity of the institution. Support for "throughput and equity" for those named as English Additional Language ("EAL") students should continue, but should be extended to "local and international" students at "key transitions" in both undergraduate and postgraduate degrees (p. 4). "Educational disadvantage" is extended to include "students from better resourced schools" (p. 2), since schooling in general is not considered as preparation for university learning. Also, differences across faculties and between "the workplace and university" (p. 3) require that disciplinary lecturers "embed academic literacy in faculties across the degree process" (p. 4).

The Plan groups the remaining three aspects of a student's repertoire illustrated in Table 1 under objective two, "Multilingualism Strategy". The claims in this section are given legitimacy by references to examples of existing practice and institutional statistics on language diversity. Firstly, "multilingualism for learning" establishes that languages other than 'English' can be used for learning in "formal" spaces. Glossaries, "multilingual study material" and "multilingual tutors" can support "concept literacy" for *epistemological access* to and *achievement* in disciplines such as "mathematics" and "statistics", and fields such as "Humanities", "Science" and "Health Sciences" that have specialised language use (p. 6).

The second multilingual strategy, to be planned at faculty level, involves all students in professional degree programmes "tak[ing] at least one semester course in an African language" (p. 7). This strategy shifts the 'language problem' to monolingual English students, but only those in professional programmes, illustrated by practice in Health Sciences. This strategy is "urgent" (p. 7) for *achievement*, given external pressure from professional and educational accreditation bodies and government. The final two multilingual strategies foreground *diversity* of named languages for *social access* on campus; "multilingualism" to "enhance social interaction" (p. 8) in "informal contexts" requires "communication" courses in 'isiXhosa' and 'Afrikaans', and "multilingualism for transforming the institutional environment" (p. 8) requires institutional communication in 'English', 'isiXhosa' and 'Afrikaans'.

While promoting these four multilingual strategies, the Plan identifies financial and human resource constraints and difficulties making "space" (p. 6) for African language courses in Health Sciences curricula. "Multilingual study material" (p. 6) should be made available online, but *power* asymmetries in *physical access* to digital technology are not surfaced.

The Plan names the third Policy objective, the development of 'isiXhosa' for learning and scholarship, as a "requirement" of national LPHE, and reproduces the national and institutional policy language of "promotion", given the financial, human and structural constraints (p. 9). Importantly, the aforementioned STEM-specific examples of glossaries and tutorials are noted as contributions to the necessary language development. Yet the Plan identifies a need for *physical access* for further development; building capacity for this work requires a structural pipeline to and funding for an undergraduate major and postgraduate study in 'African' languages.

Thus, the 2013 Plan responds to its mandate, while also developing two critical meanings of equity. University language use is *diverse,* involving a range of identities and context-specific use socially, professionally and academically, and at all levels of study. Disciplinary-specific language use is noted, but most importantly, examples presented are from STEM. Reading and digital literacies are recognised language modes, but the visual and symbolic modes for STEM learning are absent. Who needs support is extended beyond the "EAL" student, while recognising the need for ongoing support for the "EAL" student, given the *power* of 'English' in the Anglonormative space. Yet given its Policy mandate and relatively minimal detail on indigenous language intellectualisation, the Plan is limited with regards to challenging the dominant use of named languages. Hence possibilities for language and knowledge *design* are not developed.

5.3   UCT Vision and Strategy Since 2013

In the absence of an updated, published UCT language policy, we turn to more recent institutional and STEM Faculty level vision and strategy texts. We acknowledge that these texts were not developed specifically as language policy, but we are interested in references to language.

The institutional Strategic Planning Framework (UCT, 2016) for 2016 to 2020 mentions language in two of the five goals, with action not prescribed. Firstly, building "a new inclusive identity" (p. 10) is about *social access* to UCT, with the presence on campus of a *diversity* of named languages, together with different religions, cultures, political views and so on, given significance. This includes the use of indigenous languages along with attention to artworks and building names. Secondly, for "innovation in teaching and learning", language, culture and experience are "resources" which should be "recognised" and "utilised" (p. 30), a hint at their use for *epistemic access* and *design*. Monolingual 'English' students need to expand their language use; they should be "encouraged" to "acquire communicative competence" in a South African indigenous language and to learn "other major world languages", "especially" those used elsewhere in Africa (p. 31). The UCT 2030 Vision, currently a discussion document (UCT, 2020), makes one reference to language in its recognition of the institution's history as an "English-speaking colonial university". It seeks to value its "Afrikan roots" (p. 5), the intentional naming "Afrika" asserting the agency of the continent. The institution has a "dream" to draw on its "social and cultural diversity" (p. 7) and to contribute locally and globally. Thus, attention is given to geography, and not language, in identifying the university community and its relevance.

We focus next on the Faculty of Engineering and the Built Environment (EBE) (n.d.) Strategic Plan for 2017–2020, and the Faculty of Health Science (2015) "work in progress" Vision for 2030. Staff and student *diversity* for *social access* focuses variously on racial and gender categories, culture, values and epistemology. EBE identifies "multilingual" signage and letterheads, and staff participation in conversational 'isiXhosa' as necessary for "inclusivity" (p. 1). EBE suggests "differential entry targets" for students (p. 4), and the Faculty of Health Sciences wants undergraduate and postgraduate intake to "meet the needs of the country" (p. 4). This could include language, since the latter Faculty currently does recognise school credits in a named 'African' language in undergraduate admissions. Regarding the nature of STEM knowledge, curriculum in EBE should be "inclusive, relevant and contextual" (p. 4), and clinical work in Health Sciences needs a "patient-centred approach" (p. 10). Yet across these texts, the role of *diverse* language resources for learning and knowledge production is absent.

6   Conclusions and Recommendations

We began by arguing that language policy analysis for equity in STEM education requires conceptual tools that respond to the specificity of context, but which also potentially make a theoretical contribution in ways that do not universalise. Writing from a postcolonial context requires tools that recognise how notions of 'language' and 'STEM knowledge' are constructed in particular historical and geo-political relations of coloniality and global neoliberalism, and conceptualise equity as more than access to and achievement in the dominant knowledge using historically dominant languages. We have used the case of language policy at UCT to illustrate how our proposed tools bring into view how policy may enable or constrain equity in that context. To conclude, we summarise this analysis and then illustrate, in the form of recommendations, how this knowledge can inform policy development.

Our analysis suggests that the discourse on language in the 2013 UCT texts largely focuses on 'English' for *epistemic access* to and *achievement* in dominant STEM knowledge. Elite multilingualism, for which space has to be made, mainly signals *diversity* for *social access*. Yet, importantly, the 2013 Plan develops two critical meanings of equity, in particular by expanding the perspective on language, who needs language support, and how indigenous language might be used for *epistemic access* to dominant knowledge taught and assessed in 'English'. Crucially, the Plan identifies disciplinary-specific language use not only in 'English' but also in indigenous languages, exemplified in use in STEM. Yet there are limited opportunities for working with the concept of *design* in the sense of expanding what counts as language use for producing STEM knowledge.

We argue that the discursive shifts made in the 2013 Plan, ongoing on the ground STEM language practice, and the new national policy provides the push and space for a revised UCT policy that attends to all five, interrelated meanings of equity. Thus, we end with five contributions to this future policy development that might also be considered in other contexts of language diversity.

Firstly, language policy itself needs to be recognised as historical, social and political text, that works ideologically, and has material effects. So policy, and not only practice, can (re)produce or challenge inequity. This text needs to act at institutional level, but also faculty and disciplinary level, and intertwined with other policy and strategy at these multiple levels.

Secondly, language policy development can be informed by the five-part notion of equity exemplified in this chapter. Crucially, attention to the three critical meanings involves challenging the *power* of monoglossic ideologies of language by offering a heteroglossic perspective of all language use as practice that changes and develops in use and functions ontologically, epistemologically and relationally, and language as multimodal (including the visual and symbolic modes for STEM). This also involves challenging the dominance of 'English' as the language of learning and scholarship, by drawing on existing work on the intellectualisation of indigenous languages for use in STEM.

Working at the level of ideology facilitates a move from seeing language *diversity* as an elite multilingualism for *social access*, to a critical view of heteroglossic language use and indigenous multilingualism as a resource for *social* and *epistemic access* to dominant knowledge and also for *design* of STEM knowledge. Indeed, attention to *power* and *diversity* from a critical perspective shifts the definition of the 'language problem' from the student who is not 'proficient' in 'English'. For it draws attention to possibilities of *design* in the form of the related processes of 'reinventing' language in the sense that indigenous languages develop through their use (Finlayson & Madiba, 2002; Mkhize et al., 2014), expanding what counts as legitimate language use for STEM, and building new, quality meanings as relevant in a contemporary, postcolonial world (Makoni & Pennycook, 2007).

Development of the three critical meanings strengthens the concept of *access* for equity. So rather than multilingualism acting symbolically in the service of *social access*, it is viewed as acting to build both identity and knowledge for *achievement* in quality, locally and globally relevant STEM knowledge. For *physical access*, student admissions in all STEM faculties and staff recruitment and selection must value indigenous languages. Not only does this raise the status of these languages, but it creates space for these languages to develop in their use in teaching and learning, and in scholarship (Finlayson & Madiba, 2002; Mkhize et al., 2014).

Attention to staff recruitment and selection is important for our third recommendation, which is for the university to take seriously whose voices contribute to policy development (Antia & van der Merwe, 2019), both in terms of what language repertoires but also what disciplines are represented.

Fourth, we argue that policy needs to act at multiple levels. As noted, it has to act at the macro-level of ideology and at the level of STEM disciplines, drawing on language scholarship in these disciplines for legitimacy. Importantly it needs to work from the ground up with practical examples of the use of indigenous 'African' languages in STEM. Certainly, we have many promising examples of the process of intellectualisation of indigenous languages in South Africa. This includes dynamic and ongoing translation processes for glossaries in economics, mathematics and statistics (Madiba, 2014) and psychology (Mkhize et al., 2014). It includes examples of how glossaries and translanguaging practices can be integrated into university classrooms to promote both *epistemological* and *social access* in these disciplines (Madiba, 2019; Mkhize et al., 2014; Paxton, 2019; Whitelaw et al., 2019), and in computer science (Dalvit, 2010), while simultaneously furthering intellectualisation of the languages through their use. We also have examples of how attention to indigenous languages contributes to the *design* of new knowledge in astronomy (Leeuw, 2014), computer science (Dalvit, 2010) and psychology (Mkhize et al., 2014). There is also a growing body of work in South Africa focusing on multimodal STEM language use, including the visual and symbolic modes; in earth and life sciences (Paxton et al., 2017), civil engineering (Simpson, 2015) and engineering dynamics (Le Roux & Kloot, 2020).

Finally, we argue that the development of policy as discursive text as proposed here needs to be seen as acting materially with other resources, both financial and human. For example, the institution could offer collaborative education teaching and learning grants involving disciplinary and language experts working on the ground with visible policy text.

References

Antia, B. E., & van der Merwe, C. (2019). Speaking with a forked tongue about multilingualism in the language policy of a South African university. *Language Policy, 18*(3), 407–429.

Bakhtin, M. M. (1981). Discourse in the novel (C. Emerson & M. Holquist, Trans.). In M. Holquist (Ed.), *The dialogic imagination: Four essays.* Austin, TX: University of Austin Press.

Bishop, A. J. (1990). Western Mathematics: The secret weapon of cultural imperialism. *Race & Class, 32,* 51–65.

Bourdieu, P. (1991). *Language and symbolic power.* Boston, MA: Harvard University Press.

Cele, N. (2004). 'Equity of access' and 'equity of outcomes' challenged by language policy, politics and practice in South African higher education: The myth of language equality in education. *South African Journal of Higher Education, 18*(1), 38–56.

Dalvit, L. (2010). *Multilingualism and ICT education at Rhodes University: An exploratory study.* Doctoral dissertation, Rhodes University.

Department of Higher Education and Training (DHET). (2020). *The language policy framework for public higher education institutions* (Gazette No 43860). Pretoria, South Africa: DHET.

Duminy, J., Watson, V., & Odendaal, N. (2014). Introduction. In J. Duminy, J. Andreasen, F. Lerise, N. Odendaal, & V. Watson (Eds.), *Planning and the case study method in Africa: The planner in dirty shoes* (pp. 1–17). London, UK: Palgrave Macmillan.

Faculty of Engineering and the Built Environment. (n.d.). *Faculty of Engineering & the Built Environment: Strategic plan (2017–2020).* https://www.ebe.uct.ac.za/sites/default/files/image_tool/images/50/home/EBE%20STRATEGIC%20PLAN%20APPROVED.pdf.

Faculty of Health Sciences. (2015). *Strategic plan: Vision 2030 (Work in progress).* https://www.health.uct.ac.za/sites/default/files/image_tool/images/116/aboutus/Vision%202030%20January%202015%20website.pdf.

Fairclough, N. (2003). *Analysing discourse: Textual analysis in social research.* London, UK: Routledge.

Fairclough, N. (2006). *Language and globalization.* London, UK: Routledge.

Finlayson, R., & Madiba, M. (2002). The intellectualisation of the indigenous languages of South Africa: Challenges and prospects. *Current Issues in Language Planning, 3*(1), 40–61.

García, O., & Lin, A. M. (2018). English and multilingualism: A contested history. In P. Seargeant, A. Hewings, & S. Pihlaja (Eds.), *The Routledge handbook of English language studies* (pp. 77–92). London, UK: Routledge.

Gillespie, K., & Naidoo, L. (2019). #MustFall: The South African student movement and the politics of time. *The South Atlantic Quarterly, 118,* 190–194.

Glissant, É. (2010). *Poetics of relation* (B. Wing, Trans.). Anne Arbor, MI: University of Michigan Press.

Gutiérrez, R. (2002). Enabling the practice of mathematics teachers in context: Toward a new equity research agenda. *Mathematical Thinking and Learning, 4*(2–3), 145–187.

Gutiérrez, R. (2012). Context matters: How should we conceptualize equity in mathematics education? In B. Herbel-Eisenmann, J. Choppin, D. Wagner, & D. Pimm (Eds.), *Equity in discourse for mathematics education: Theories, practices, and policies* (pp. 17–33). Dordrecht, The Netherlands: Springer.

Gutiérrez, R., & Dixon-Román, E. (2011). Beyond gap gazing: How can thinking about education comprehensively help us (re)envision mathematics education? In B. Atweh, M. Graven, W. Secada, & P. Valero (Eds.), *Mapping equity and quality in mathematics education* (pp. 21–34). Dordrecht, The Netherlands: Springer.

Janks, H. (2010). *Literacy and power*. New York, NY: Routledge.

Kapp, R., & Arend, M. (2011). 'There's a hippo on my stoep': Constructions of English second language teaching and learners in the new National Senior Certificate. *Per Linguam, 27*(1), 1–10.

Kotzé, E. (2014). The emergence of a favourable policy landscape. In L. Hibbert & C. van der Walt (Eds.), *Multilingual universities in South Africa: Reflecting society in higher education* (pp. 15–27). Bristol, UK: Multilingual Matters.

Kuteeva, M., & Airey, J. (2014). Disciplinary differences in the use of English in higher education: Reflections on recent language policy developments. *Higher Education, 67*(5), 533–549.

Leeuw, L. L. (2014). An exemplary astronomical lesson that could potentially show the benefits of multilingual content and language in higher education. In L. Hibbert & C. van der Walt (Eds.), *Multilingual universities in South Africa: Reflecting society in higher education* (pp. 167–178). Bristol, UK: Multilingual Matters.

Le Roux, K., & Rughubar-Reddy, S. (forthcoming). The potential of an Africa-centred approach to theory use in critical mathematics education. In R. Barwell & A. Andersson (Eds.), *Applying critical mathematics education*. Brill.

Le Roux, K., & Kloot, B. (2020). Pedagogy for modelling problem solving in engineering dynamics: A social semiotic analysis of a lecturer's multimodal language use. *European Journal of Engineering Education, 45*(4), 631–652.

Luckett, K. (2016). Curriculum contestation in a post-colonial context: A view from the South. *Teaching in Higher Education, 21*(4), 415–428.

Madiba, M. (2014). Promoting concept literacy through multilingual glossaries: A translanguaging approach. In L. Hibbert & C. van der Walt (Eds.), *Multilingual universities in South Africa: Reflecting society in higher education* (pp. 68–87). Bristol, UK: Multilingual Matters.

Madiba, M. (2019, March 29). *Teaching multilingual and multicultural students: Translanguaging as an innovative pedagogy.* Plenary address at the GFD Symposium 2019 | Frontiers in Higher Education: Diversifying and Transforming Teaching, University of Tokyo, Japan. https://www.gfd.c.u-tokyo.ac.jp/albums/abm.php?f=abm00022691.pdf&n=Madiba_GFD+Symposium_Keynote.pdf.

Makoe, P., & McKinney, C. (2014). Linguistic ideologies in multilingual South African suburban schools. *Journal of Multilingual and Multicultural Development, 35*(7), 658–673.

Makoni, S., & Pennycook, A. (2007). Disinventing and reconstituting languages. In S. Makoni & A. Pennycook (Eds.), *Disinventing and reconstituting languages* (pp. 1–41). Clevedon, UK: Multilingual Matters.

Mbembe, A. J. (2016). Decolonizing the university: New directions. *Arts and Humanities in Higher Education, 15*(1), 29–45.

McKinney, C. (2017). *Language and power in post-colonial schooling: Ideologies in practice.* New York, NY: Routledge.

Ministry of Education. (2002). *Language policy for higher education.* South Africa: Pretoria.

Mkhize, N., Dumisa, N., & Chitindingu, E. (2014). Democratising access and success: IsiZulu terminology development and bilingual instruction in psychology at the University of KwaZulu-Natal. *Alternation Special Edition, 13,* 128–154.

Morrow, W. (2009). *Bounds of democracy: Epistemological access to higher education.* Cape Town, South Africa: HSRC Press.

Nudelman, C. (2015). *Language in South Africa's higher education transformation: A study of language policies at four universities.* Masters dissertation, University of Cape Town.

Pais, A. (2012). A critical approach to equity. In O. Skovsmose & B. Greer (Eds.), *Opening the cage? Critical agency in the face of uncertainty* (pp. 49–91). Rotterdam, The Netherlands: Sense.

Paxton, M. I. J. (2009). 'It's easy to learn when you using your home language but with English you need to start learning language before you get to the concept': Bilingual concept development in an English medium university in South Africa. *Journal of Multilingual and Multicultural Development, 30*(4), 345–359.

Paxton, M., Frith, V., Kelly-Laubscher, R., Muna, N., & van der Merwe, M. (2017). Supporting the teaching of the visual literacies in the earth and life sciences in higher education. *Higher Education Research & Development, 36*(6), 1264–1279.

Prah, K. K. (2017). The intellectualisation of African languages for higher education. *Alternation, 24*(2), 215–225.

Rooney, C. (2015). *Using survival analysis to identify the determinants of academic exclusion and graduation in three faculties at UCT.* Masters Dissertation, University of Cape Town.

Shohamy, E. G. (2006). *Language policy: Hidden agendas and new approaches.* Abingdon, UK: Psychology Press.

Simpson, Z. S. (2015). *Students' navigation of multimodal meaning-making practices in civil engineering: An (auto)ethnographic approach.* Doctoral thesis, University of Cape Town.

Skovsmose, O. (2016). Mathematics: A critical rationality? In P. Ernest, B. Sriraman, & N. Ernest (Eds.), *Critical Mathematics education: Theory, praxis and reality* (pp. 1–22). Charlotte, NC: Information Age.

Statistics South Africa. (2019). *General household survey 2018.* Pretoria, South Africa: Statistics South Africa.

Tyler, R. (2016). Discourse-shifting practices of a teacher and learning facilitator in a bilingual mathematics classroom. *Per Linguam, 32*(3), 13–27.

Van der Merwe, C. (2016). *Analyzing university language policies in South Africa: Critical discourse and policy analysis frameworks.* Masters Dissertation, University of the Western Cape.

University of Cape Town. (2013a). *Language policy.* https://www.uct.ac.za/sites/default/files/image_tool/images/328/about/policies/Language_Policy_19-June-2013_Final.pdf.

University of Cape Town. (2013b). *UCT language policy implementation plan* (Draft). University of Cape Town, South Africa.

University of Cape Town. (2016). *Strategic planning framework.* https://www.paperturn-view.com/newsroom-and-publications/strategic-plan-digimag-v2?pid=MjA20459.

University of Cape Town. (2020). *Vision 2030 discussion document.* https://www.news.uct.ac.za/images/userfiles/files/publications/VC_Vision.pdf.

Whitelaw, E., Dowling, T., & Filby, S. (2019). Leveraging language: Preliminary evidence from a language-based intervention at the University of Cape Town. *Critical Studies in Teaching and Learning, 7*(2), 75–93.

Kate le Roux is Associate Professor in Language Development in the Centre for Higher Education Development, University of Cape Town, South Africa. Her research is located at the intersection of language, mathematics and the learning of disciplinary knowledge in science and engineering. This focuses on equity, power and identity, drawing theoretically from critical linguistics, critical mathematics education, multilingualism and multimodality for learning, and Southern Theory. She was a 2014 Mandela Mellon Fellow at the Hutchins Center for African & African American Research, Harvard University, USA.

Bongi Bangeni is Associate Professor in Language Development in the Centre for Higher Education Development, University of Cape Town, South Africa, and a Mandela Fellow at the Hutchins Center for African & African American Research, Harvard University, USA. She has published on writing and identity, black students' language attitudes and their transitions from undergraduate to postgraduate studies. She is co-editor of the book Negotiating Learning and Identity in Higher Education: Access, Persistence and Retention (2017, Bloomsbury) which forms part of the 'Understanding Student Experiences of Higher Education' series.

Carolyn McKinney is Associate Professor in Language Education at the University of Cape Town, South Africa. Carolyn's teaching and research focuses on language ideologies; multilingualism as a resource for learning; critical literacy and the relationships between language, identity/subjectivity and learning. She recently published Language and Power in Post-Colonial Schooling: Ideologies in Practice (2017, Routledge) and is a member of the bua-lit language and literacy collective: http://www.bua-lit.org.za/.

The Place Where Languages Meet to Argue: A Contribution from an Analysis of the Brazilian National Curriculum

Renata de Paula Orofino, Nathália Helena Azevedo, and Daniela Lopes Scarpa

Abstract Argumentation has long been established by science education research as the language of sciences. While argumentation is central to science education, language curricula are also responsible for teaching argumentation. Therefore, there is a potential for an interdisciplinary approach to argumentation. Brazilian educational policies explicitly address argumentation under the idea of both integral education and interdisciplinarity. This work aimed at discussing how argumentation is conceived in the Brazilian National curriculum and how science teachers may act upon the curriculum and benefit from an interdisciplinary approach to teach argumentation as the process of constructing scientifically sound arguments. Our main findings are that there is a huge focus on argumentation in the Natural Sciences curriculum, that the curriculum focuses mainly on skills of construction of arguments and its parts, without addressing what arguments are and how to use language to argue. Arguments in Natural Sciences are supported by evidence, but there is no room for alternative conclusions from the same data set nor rhetorical aspects. The disciplines are not connected in the Brazilian main curricular document, contradicting the same policy that argues for integral and interdisciplinary education. Brazilian curriculum leaves for the teachers to figure out how to work argumentation connecting languages and Natural Sciences and may even hinder teachers' work to reach interdisciplinarity.

Keywords Argumentation skills · Science education · Educational policies · Curriculum

R. de Paula Orofino (✉)
Federal University of ABC, Santo André, Brazil
e-mail: renata.orofino@ufabc.edu.br

N. H. Azevedo · D. L. Scarpa
University of São Paulo, São Paulo, Brazil
e-mail: helena.nathalia@usp.br

D. L. Scarpa
e-mail: dlscarpa@usp.br

© The Author(s), under exclusive license to Springer Nature Switzerland AG 2021
A. A. Essien and A. Msimanga (eds.), *Multilingual Education Yearbook 2021*, Multilingual Education Yearbook,
https://doi.org/10.1007/978-3-030-72009-4_12

1   Introduction

In this chapter, we aim at understanding how argumentation is conceived across the Brazilian National curriculum and the possibilities for science teachers to act on the curriculum and teach argumentation articulating an interdisciplinary work. Argumentation is relevant for science education and we focus on the opportunities given by Brazilian educational policies for teaching sciences as argument (Kuhn, 1993). Languages (i.e. English and Portuguese) are also subjects in the curriculum responsible for argumentation education. Natural Sciences teachers can benefit from understanding how each subject tackles argumentation and analysing the curricular document may help us find the spaces for it.

Argumentation takes part in different social contexts. Science as a culture has argumentation as one of its key linguistic aspects. One could consider argumentation the language of Science since when engaging in epistemic practices, one uses argumentation as the main 'language' (Kuhn, 1993; Kelly & Licona, 2018). Arguing in scientific contexts also involves understanding other facets of the language, in the sense of acquiring a particular idiom. Science education can benefit from language education when teaching argumentation.

Argument-based learning contributes to problem-solving and the use of critical thinking in science education (e.g. Kuhn et al., 2017), skills required in addressing STEM (sciences, technology, engineering, and mathematics)[1] education. Additionally, by engaging youth in inquiry and problem-based learning, we foster epistemic practices of STEM such as argumentation, and promote different degrees of integration of STEM (e.g. Kelly & Knowles, 2016). The STEM education movement has become a global education policy agenda (Mizell & Brown, 2016) and, despite its multiple meanings, studies on STEM echoes on interdisciplinarity, and on the benefits of STEM approach to scientific literacy (Martín-Páez et al., 2019).

Brazilian educational policies are organized in a set of documents (a law, guidelines, a ten years' plan, and a curricular base) with tenets that guide educational goals in the country. One of the tenets is that the student must be able to integrally comprehend "the natural and social environment, the political system, the technology, the arts and the values in which society lies", expressed in the Law of Directives and Basis for Brazilian Education (LDB in Portuguese, BRASIL, 1996). The concept of integral education was first used in Brazil in the 1930s by the Pioneers of Education Movement, in reference to the importance of the social aspects of education. This means that education is committed to respecting the complexity of human development and growth, aiming for more than just the cognitive or intellectual dimension of education (MEC, 2018). The National Curricular Guidelines for Basic Education (DCN in Portuguese, BRASIL, 2013) highlights that the curricular components must be organized in areas, disciplines, and thematic axis, preserving the particularities of

---

[1] We acknowledge the broader discussion concerning STEM education. This chapter does not participate directly in such discussion due to the particularities of the analysed context. Still, there are aspects of Brazilian educational policy that allow us to argue for the relevance of the analysis put forth in this chapter.

the different fields of knowledge, in a rhythm compatible to the integral development of the citizen. This same document argues for transcending the fragmentation of the areas, aiming for an integration of the curriculum that allows students with different skills, life experiences, and interests to participate in learning. The Common National Curricular Base (BNCC in Portuguese; MEC, 2018) is the most recent document and it is supposed to rule the construction of states' and districts' curricula, teacher education, the production of pedagogical resources, and national assessments. The BNCC reinstates the integral education tenet and recalls DCN's ethical, political, and aesthetical principles (BRASIL, 2013).

Another tenet of Brazilian educational policies worth mentioning is interdisciplinarity. Specially regarding high school, LDB and DCN highlight technology in all areas of knowledge for basic education (e.g. Natural Sciences and its technologies). Even though BNCC does not use the term STEM education, it sets seven general competencies that imply the integration of the disciplines:

> Exercise intellectual curiosity and recur to sciences' own approach, including the investigation, the reflection, the critical analysis, the imagination, and the creativity to investigate causes, elaborate and test hypotheses, formulate and solve problems, and create solutions (including technological ones) based on the knowledge of the different areas. (MEC, 2018, p. 9)

> Use different languages - verbal, corporal, visual, sonorous, and digital - as well the knowledge of artistic mathematical and scientific languages to express and share information, experiences, ideas, and feelings in different contexts and make sense that aims at mutual understanding. (MEC, 2018, p. 9)

> Argue based on reliable facts, data, and information in order to articulate, negotiate and support common ideas, points of view and decisions showing respect to and promoting human rights, socio environmental awareness and sustainable consumption locally, regionally and globally, ethically positioning oneself and taking care of oneself, the others and the planet. (MEC, 2018, p. 9)

Thus, inquiry, scientific language, and argumentation are considered as important competencies to be developed throughout all the curricular components following the integral education context of Brazilian educational policies. We chose to analyse how argumentation is conceived across Natural Sciences, Portuguese, and English national curricula and the possibilities for science teachers to act on the curriculum and teach argumentation articulating an interdisciplinary work in order to discuss the possibilities of complementarity and interdisciplinarity between each curriculum regarding argumentation.

2 Theoretical Framework

Science education research addresses the central role of argumentation in science education since the end of the 1990s (Lee et al., 2009). Kuhn (1991), after investigating 'thinking as argument' in her research on people's arguments about social issues, acted as one of many inspirations to Driver et al.'s paper (2000). Such paper became the most cited paper in the early 2000s in science education research (Lee et al., 2009) and was extremely relevant to further research on argumentation.

Argumentation is well established as the typical scientific discourse (Kuhn, 1993), but we also got to know how argumentation plays in learning scientific concepts (Zohar & Nemet, 2002; Ryu & Sandoval, 2012), in understanding Science as a culture (Jiménez-Aleixandre & Erduran, 2008), in developing reasoning skills, critical thinking, and metacognition (Lin et al., 2014), argumentation as an epistemic practice (Bricker & Bell, 2008), and on the importance of arguing when learning socio-scientific issues (Sadler, 2004). Argumentation can also be considered as a form of cultural hybridization, which makes it possible to connect school and Science's cultures (Scarpa & Trivelato, 2013).

Much has been researched on linguistic aspects of argumentation (Erduran et al., 2015), on rhetorical skills and persuasion (e.g. Mendonça & Justi, 2013), on how teachers can foster students to argue in science education (e.g. Simon et al., 2006), and on the nuances of oral and written argumentation (e.g. Kelly & Takao, 2002; Zohar & Nemet, 2002). For all these aspects and inspired by Kuhn's ideas (1993, 2010), we approach argumentation as a 'language', the language of sciences.

The basic ideas of argumentation from a philosophical perspective (i.e. premises and conclusions form an argument) are already interesting for argumentation analysis in science education. More importantly, science education research's main goal when analysing arguments is different from the formal logical one. When investigating argumentation, science education research has relied largely on Toulmin's Argument Pattern (TAP; Toulmin, 1958), other informal logic frameworks (e.g. Walton, 1996), and pragma-dialectic perspectives (Van Eemeren et al., 2002). Either way, all analytical frameworks so far link argumentative discourse with scientific language. In this paper, we draw from an informal logic approach to argumentation (Toulmin, 1958; Walton, 1996). Also, we draw from the science education literature on how to teach argumentation (Kuhn, 1993, 2010; Simon et al., 2006).

Toulmin (1958) and Walton (1996), each in its own way, defied formal logic in the mid-1900s and created space for argumentation analysis to include day to day argumentation. The goal of argumentation shifted from specific sets of rules to the need for people to keep the conversation going when building knowledge.

Toulmin provided us with the idea that there are different roles for each premise in the argument, that we do not always need one major and one minor premise in order to have valid arguments, and that what formal logic calls valid can be distinguished in several forms of argument (Velasco, 2009). In TAP, whatever the nature of a particular assertion (Claim), it can be challenged about its grounds (Data, Warrants, and Backings). Data are the facts that support the Claim, while Warrants are the bridges that legitimate the Claim. Warrants are general statements based on specific rules and laws according to the field of discussion (Backings). Warrant's strength makes the argument probable or certain (Qualifier), while Rebuttal indicates the situations that invalidate the argument (Toulmin, 1958).

Walton (1996), on the other hand, used classical argumentation schemes to defend that not always a fallacy should be considered as such. Walton indicates that presumptive inferences have been ignored by logicians, but this kind of inference allows the dialogue to go ahead on a provisional basis and provides a tentative solution to practical. From this point of view, the circumstances of a particular case are taken into account before stating that an argument is fallacious (Walton, 1996). With his argumentation schemes and critical questions, Walton helped us understand that abductive and presumptive arguments may be crucial to keep the dialogue going until we can test hypothetical ideas, central in scientific contexts.

When it comes to science education research, Kuhn (1993) addressed the importance of linking the ways scientists and ordinary children think as a way to reconnect school children to scientific subjects. From Kuhn's perspective, scientific facts as well as scientific theories are argumentative constructions, which means that argumentation is core to scientific enterprise and education. Kuhn draws from informal reasoning with a dialogical perspective of argumentation.

Simon et al. (2006) contributed to science education research on argumentation because they implied that there are actions teachers may endure fostering students' argumentation. Their exploratory work on how teachers developed argumentation activities helped us understand that the mere act of asking a student to listen to the other is important to teach how to argue. Simon et al. (2006) used TAP as their approach to argumentation.

Argumentation is a key element in improving scientific literacy (Driver et al., 2000), and is consensually one of science education goals for science education researchers. Scientific literacy is a broad concept that can be summarized as: (i) learning sciences; (ii) learning about sciences; (iii) doing sciences; (iv) addressing socio-scientific issues (Hodson, 2014). While exploring their understanding in sciences classes, fostered by teacher's planned activities (Hodson, 2014), students become aware and appropriate the typical ways-of-being, the cultural practices, and the discourse of sciences (Brown et al., 2005). That said, a science education curriculum aligned with scientific literacy goals should portray or be centered in argumentation. Sciences rely on argumentative discourse to both build conclusions about the natural world and make science accessible for public assessment.

Although STEM education is being implemented in different contexts using different teaching strategies and methods (Dass, 2015)—driven even by educational technologies—the emphasis on argumentation-based learning can be a valuable strategy for STEM learning goals, given the evidence of its effectiveness in learning complex topics in recent years. Not only argumentation should be central in curricula aligned to scientific literacy, learning how to argue can help develop a range of skills required in STEM careers (Mathis et al., 2017) and should participate in STEM curricula.

Argumentation is also key to different daily interactions in each specific language (in Brazil's case, Portuguese). Although reasoning skills can be used in different ways depending on the cultural context, as indicated by studies on arguing in intercultural contexts (e.g. Dolina & Cecchetto, 1998), there is evidence that the reasoning skills involved in arguing can be universal (Mercier, 2011). As examples, we can highlight the role of argumentation for the development of communication and interactions, its potential to contribute to social performance, and its role in building people's identities.

Hand in hand with argumentation, reading skills, and proficiency are tied with sciences and STEM education. Brazil's proficiency scores in reading are low (OECD, 2019). Research has shown that students that struggle with reading in their native language also struggle with scientific language (Fang, 2006). Teaching as argument is, therefore, broader than science education and can shape language education.

We discuss curriculum from Gimeno-Sacristán's (2017) perspective, by which curriculum exists in several dimensions: (i) prescribed curriculum (designed by governmental institutions and authorities of the educational system); (ii) curriculum posed to teachers (depicted in the textbooks); (iii) curriculum shaped by teachers (professed in teaching planning); (iv) curriculum in action (impersonated in the pedagogical practice); (v) curriculum accomplished (the effects of the educational process); and (vi) curriculum evaluated (demanded in tests and assessments). In this paper, we are interested in looking at the dimension of the prescribed curriculum, but without neglecting the fact that conceiving argumentation teaching permeates all of these different dimensions.

Somehow, when interpreted by teachers, the prescribed curriculum is transformed by teachers' knowledge of their educational contexts, thus the curriculum's original intentions may be enhanced or weakened (Gimeno-Sacristán, 2017). It is this set of changes that allows a view of the curriculum as a source of experience. The analysis of the curriculum focused on the document's expressed meanings, perspectives, and potential links is relevant since they can be used as in-service education materials, broadening the connections between theoretical and pedagogical perspectives and contributing to the growth of conscience of teachers' agency towards curricular propositions. Curriculum analysis is a facet of educational research. There are limits to what can be put in the curriculum. Yet, what is in the curriculum is what its authors chose as worthwhile and necessary, an 'ideological terrain' (Cohen et al., 2011). Our analysis in this chapter concerns analysing Brazilian policies and their implications for teaching sciences as argument (Kuhn, 1993). Also, we aim at promoting space to discuss how argumentation in languages curricula may inform different contexts of STEM teaching and learning, including multilingual ones.

3 Methodology

3.1 *Brazilian Common National Curricular Base (BNCC)*

The BNCC (MEC, 2018) is the main curricular document in Brazilian educational policy. Since BNCC prescribes general and specific competencies, and skills to each curricular component, it is an interesting resource to be analysed and helps us understand how Natural Sciences teachers may benefit from learning how argumentation can be explored and taught throughout disciplines, fulfilling the integral education tenet of Brazilian policy. The document is organized in 'initial years', from 1st to 5th year (equivalent to elementary school) and 'final years', from 6th to 9th years.

There are five 'areas of knowledge' (Languages, Mathematics, Humanities, Natural Sciences, and Religious Education) that are subdivided into disciplines. Languages, for example, are divided into Portuguese (Brazil's official language), Arts, Sports, and English. English is considered a pre-eminent international language, dissociating it from the notion of belonging to a particular territory or typical culture hence English is in Brazilian curriculum between 6th and 9th years.

Each discipline has specific skills and learning outcomes—assertions composed of verbs that indicate the cognitive processes students will endure and the knowledge students should learn—that are indicated by codes, referring to the school year, discipline, and skill number. Since BNCC is a learning outcomes oriented curriculum, its focus is on students' needs and the intricate connection established between contents, skills, values, and attitudes, instead of on learning disciplines' traditional contents (UNESCO, 2013). It is also interesting to recall that BNCC points towards integration between disciplines, as shown earlier, by stating general competencies.

3.2 Analytical Procedures

We chose to focus the analysis on ages 6–14 (school years 1st–9th) of BNCC because in this first stage of education, students are immersed in various languages, including the scientific one. We chose to analyse both languages (Portuguese and English) and Natural Sciences curricula. We considered languages and Natural Sciences to be responsible for teaching argumentation and that these disciplines' curricula could combine skills towards this goal.

For the analysis of Portuguese and English, we divided the texts and one author read the Portuguese curriculum, while the other read the English curriculum. The reading was individual and comprised a scan through the text for information about argumentation in each language followed by a preliminary grouping of similar contents. After reading, the authors presented each other their reasoning in grouping the skills and then defined the categories. After that, all authors read the Natural Sciences curriculum and applied the categories to the text.

From the initial analysis discussion, we formulated the argumentation categories (a posteriori). All authors are familiar with argumentation papers and the main insights used came from informal logic (Toulmin, 1958; Walton, 1996) and science education literature (Kuhn 1993, 2010; Simon et al., 2006). That said, we considered explicit references for arguments' elements (i.e. premises, conclusions, inference, and counterargument) and common-sense words connected to argumentation (i.e. justifying, defending a point of view, and using evidence to back assertions).

We aligned with qualitative research and took an interpretive paradigm by trying to get inside the subject of analysis and to understand it from within (Cohen et al., 2011). The analysis and the patterns were built inductively, with a main attempt to understand Brazilian curricular reality and allow comparisons with other realities (Cohen et al., 2011). We conducted a content analysis aiming, hence, to identify categories and units of analysis that could reflect our theoretical assumptions (Anderson & Arsenault, 2001) on the nature of the curriculum.

The categorisation of the analysis units was done iteratively (Ezzy, 2002). Reflecting the natural circular process of qualitative research, the final categories (presented in the following section) were reconsidered along the validation steps among the authors. We focused mainly on the verbs of the skills to create the categories and allocate the skills in them. After creating the categories, we arranged them in a hierarchical order of complexity, using the verbs and the categories as cue to the complexity of the skills inspired by Simon et al. (2006) analysis. Then we analysed the complexity of argumentation skills in each school year and each discipline. We compared the verbs used in each category to identify any overlapping between the categories.

We also analysed the texts in the introduction of each BNCC's section (e.g. each school year) to the skills in the curriculum. We considered these introductory texts as contextual and as a parameter to the proposed skills. The analysis of the context not only materializes our goal of understanding curricular elements from within, but also serves as a source of triangulation for the categories created. All the analyses were debated until the authors reached a consensus about each categorization.

4 Findings

4.1 Exploratory and Categories

We found skills in all three disciplines compatible with argumentation skills and arranged them in eight different categories (Table 1). A small number of the curricula skills were classified in more than one category (one in Natural Sciences; four in Portuguese). That happened mostly because some of the skills had multiple verbs, leading to more than one category.

All school years had argumentation skills, in different proportions. In the Portuguese curriculum, we identified 47 in a total of 391 skills (12%), distributed in all eight categories. Some of the skills targeted more than one school year, and most argumentation skills were concentrated between 6th to 9th years ($n = 13$) and 8th to 9th years ($n = 12$).

In the English curriculum, we identified seven in a total of 88 skills (8%). The skills were in four categories (codes I, II, V, and VII), all concentrated in 9th year.

In the Natural Sciences curriculum, we identified 65 in a total of 111 skills (59%), concentrated in four categories (codes III, IV, V, and VI). School 7th year had more argumentation skills, while 2nd and 3rd years had fewer skills (Fig. 1).

Table 1 Codes and description of the categories created from the curriculum argumentation skills analysed

Code	Category	Description of category
I	Identification of one or more arguments (Premises + Conclusion)	Identification of textual parts or sentences that, together, set an argument, even without detailing the names
II	Use of argumentative textual marks	Identification and construction textual elements typical of arguments such as conjunctions, rhetorical words, or modal qualifiers
III	Construction of premises	Construction of assertions that work as premises of arguments, even without detailing the names. Can involve taking notes of observations, separating empirical from theoretical knowledge, construction justifications, selecting relevant information
IV	Construction of conclusions	Construction of assertions that work as conclusions of arguments, even without detailing the names
V	Construction of one or more arguments (Premises + Conclusion)	Construction of sentences that, together, form an argument
VI	Evaluation of an argument or its parts	Logical, material, or rhetorical evaluation of premises, conclusions, or arguments. Uses words such as analyse, evaluate, ask for clarification, adequate use, compare, recognize
VII	Rhetorical	Actions with the intention of convincing or persuading the other using arguments. It can involve identifying, constructing, or evaluating arguments, but the focus is the persuasion that derives from it
VIII	Social rules of arguing	Organization, plan, mediation of contexts in which argumentation typically occurs. That includes: debates or written activities

4.2 Arguing in Each Discipline

The only category all three disciplines had in common was 'Construction of one or more arguments' (Fig. 2). Although Natural Sciences had categories of construction, it lacked identification, use of argumentative textual marks, and rhetorical categories.

The category 'Construction of premises' was much more frequent in Natural Sciences than in Portuguese. While in Portuguese, premises emerged from notes and observations, in Natural Sciences, premises were theoretical and empirical, the latter drawn by observation, experimentation, analysis, and variables:

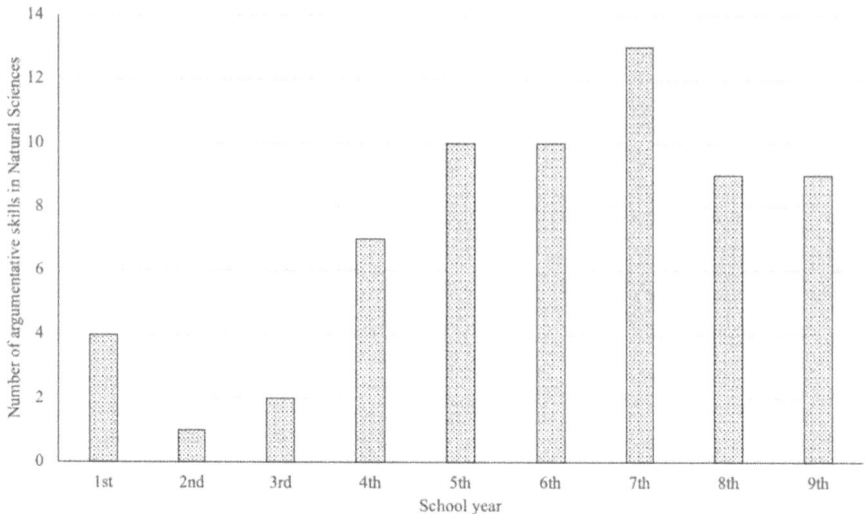

Fig. 1 Number of skills that were considered argumentative in the Natural Sciences curriculum in each school year

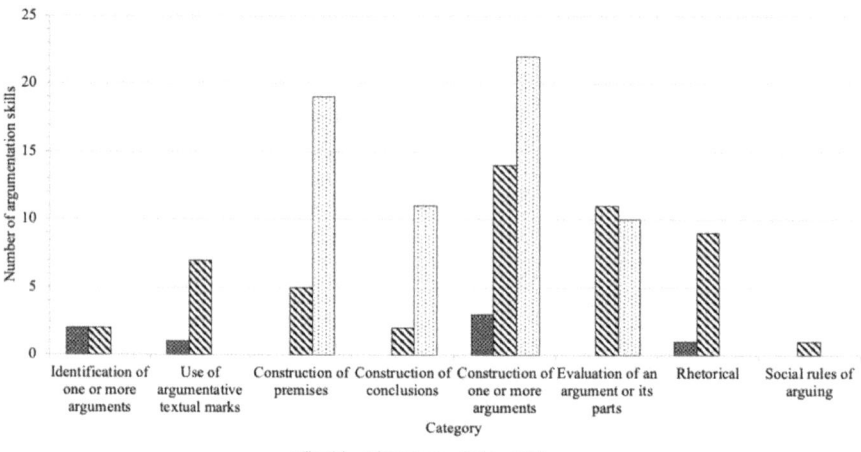

Fig. 2 Number of argumentation skills grouped by category for all three disciplines in the Brazilian curriculum analysed

Plan and write texts to present results from observations and research on information sources, including, when pertinent, images, diagrams, and simple graphs or tables, considering the communicative situation and the theme/subject of the text. (3rd year; Portuguese; MEC, 2018, p. 129)

Create different sounds from various objects' vibrations and identify the variables influencing this phenomenon. (3rd year; Natural Sciences; MEC, 2018, p. 337)

Justify the shape of Brazilian and African coast based on continental drift theory. (7th year; Natural Sciences; MEC, 2018, p. 347)

'Evaluation of arguments' appeared both in Portuguese and Natural Sciences. In the Portuguese curriculum, evaluation referred to the parts of the arguments, which means, evaluating if the premises were relevant or coherent to the conclusion, if the argument was logically valid, identifying and positioning oneself towards someone else's arguments. On the other hand, in Natural Sciences, the evaluation focused on material aspects of the arguments, which means that the skill was about selecting the best evidence to justify an assertion or evaluating the implications of a particular claim:

Analyse the validity and strength of arguments when arguing about media artefacts targeting children (movies, cartoons, comic books, games, etc.), using knowledge about those kinds of shows. (5th year; Portuguese; MEC, 2018, p. 127)

Select arguments about the viability of humanity's survival out of Earth, based on necessary conditions for life, planet's conditions, and on the distance and time of interplanetary or interstellar travelling. (9th year; Natural Sciences; MEC, 2018, p. 351)

4.3 Complexity and Cognitive Level of Argumentation Skills

As mentioned previously, we ranked the categories from simple to complex. An implicit consequence of our coding was to expect simpler categories to be more frequent in the early years and more complex categories to appear from the middle to the end of school years. That did not occur in the curricula analysed. Even with a higher number of argumentation skills in later years, Portuguese and Natural Sciences curricula had a higher number of 'Construction of one or more arguments' and 'Evaluation of an argument or its parts' as the most frequent categories (Fig. 3). The total amount of skills was higher in later years, and, in the Portuguese curriculum, all skills were present in the 8th and 9th year, with similar distribution.

We considered 'Identification of one or more arguments (Premises + Conclusion)' as the simplest category. The verbs used in these skills were mostly 'identify' and 'distinguish'. Yet, skills in this category were not present in earlier years in none of the curricula and it was only found in Portuguese (3rd to 5th and 6th to 9th years) and English (9th year) curricula:

Identify the text's main idea, demonstrating a global comprehension of it. (3rd to 5th year; Portuguese; MEC, 2018, p. 113)

Rhetorical skills were also present only in Portuguese and English curricula. Portuguese curriculum had rhetorical skills since the 3rd year, but the majority of rhetorical skills appeared from 5th to 9th years. This category was considered more complex and involved multimodal actions, such as intonation, facial and body language, word choice, and persuasion:

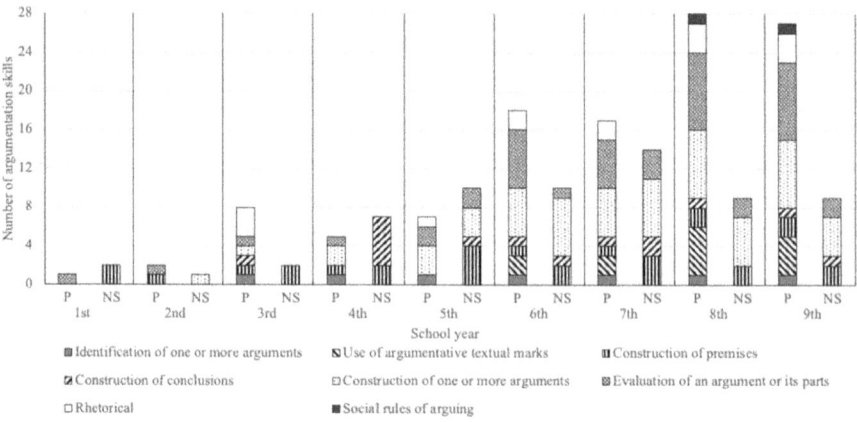

Fig. 3 Number of argumentation skills grouped by category and arranged by school year for Portuguese (P) and Natural Sciences (NS) curricula

> Identify the use of persuasive tools in various argumentative texts (as title planning, lexical choices, metaphors, expliciting or hiding sources of information) and understand its effect on text sensemaking. (6th to 5th year; Portuguese; MEC, 2018, p. 163)

Portuguese curriculum had 'Construction of one or more arguments' from 3rd year on, more frequently after 6th year, while in Natural Sciences curriculum it appeared in 2nd and in 5th year on. Portuguese curriculum stated ideas of counter arguing and respecting different points of view while the Natural Sciences curriculum did not. Even when the argument was related to innovations, solutions, or social actions, the possibility of having more than one solution or even alternative or contradictory solutions was not clear in the Natural Sciences curriculum. It can be understood as if an argument based on some evidence cannot have alternative conclusions:

> Position oneself in a consistent and supported fashion in discussions, assemblies, school's board meetings, students' union, and other situations of proposing and defending opinions, respecting other peoples' contrary opinions and alternative proposal, and backing one's positioning, using the time available, with synthetic, clear, and justified proposals. (6th to 9th year; Portuguese; MEC, 2018, p. 149)

> Propose collective actions to enhance the use of electricity in one's schools and/or community, based on the selection of equipment following sustainability criteria (energy consumptions and efficiency) and responsible consumption habits. (8th year; Natural Sciences; MEC, 2018, p. 349)

> Use the knowledge of heat propagation forms to justify the use of specific materials (conductive and isolators) in daily life; explain the functioning principles of some equipment (thermic bottles, solar panels, etc.) and/or building technological solutions applying this knowledge. (7th year; Natural Sciences; MEC, 2018, p. 347)

One peculiar finding was that there is some polysemy in the verbs used, which means that they were interpreted as different categories depending on the context of use. For example, the verb 'discuss', in the Natural Sciences curriculum appeared in 'Construction of premises', 'Construction of conclusion', and 'Evaluation of an argument or its parts'. The verb 'discuss' also appeared in skills in which discussing was synonymous with observing or noticing something.

4.4 The Social Meaning of Arguing in Each Discipline

Each curriculum had a contextual text with general guidelines for how argumentation is to be seen in each discipline. Portuguese and English were gathered in a broader introductory text called Languages. It presented the importance of negotiating and exposing points of view in oral and written actions, and in the technological world. It also asserted the importance of respecting other people's beliefs and the necessity of promoting human rights and raising critical citizens in the contemporary world:

> Using different languages to support points of view showing respect to and promoting human rights, socio-environmental awareness, and sustainable consumption locally, regionally, and globally, acting critically towards contemporary world issues. (MEC, 2018, p. 63)

English did not have any other reference to argumentation. Portuguese, on the other hand, was very rich, with links to language that can interfere in argumentation and also explicit links to argumentation. We found the recommendation that students should understand how languages are dynamic and how everyone participates at the same time in the process of transforming the language. When talking about older school years, how students should see themselves as agents in public/social life. Finally, the text brings the importance of checking sources of information and being critical towards online content, since there are reliable and unreliable websites on the internet:

> Dialogy and relationship between texts: establish intertextuality and interdiscursivity relations that allow identifying and comprehending different positionings and/or perspectives in stake, the role of paraphrase and parody and styling kinds of productions. (MEC, 2018, p. 71)

The explicit links in Portuguese contextual text comprise: (i) the influence of post-truth in one's judgement towards information; (ii) the importance of organizing information in the text to portray correctly cause and effect; and (iii) thesis, arguments, problems, and solutions, definitions, and examples. The text also brings rhetoric as a tool for promoting sustainable consumption, and dialogue, as it declares how debates and discussions can and should be used to make different voices to be heard and to value different points of view.

Natural Sciences contextual text was also very rich. It reaffirmed the idea of reporting inquiry results systematically, orally, and using multimodality. It had signs of scientific literacy when talking about multiple daily situations in which sciences are involved. Science teaching was also described as a means to both feeling comfortable when discussing scientific, technological, and environmental issues, and aiming at building social justice, democracy, and inclusion. Although the text indicated that the skills' complexity increases throughout school years, we did not see this happen with argumentation skills, as pointed earlier.

Some of the ideas brought up in the Natural Sciences contextual text did not translate into skills later in the document. A very crucial idea for arguing in sciences is the possibility and acceptance of different points of view, and the text initially values that. Still, we did not find any reference to counter arguments, rebuttals, or debates settings in the skills. Similarly, the contextual text indicates the importance of reviewing, criticizing one's own conclusions and inquiry processes, but no skill brings that idea. The text suggests that scientific ways of explaining natural phenomena change historically, but no skill indicates how argumentation can be used to contemplate and to build upon historical knowledge by reviewing models and explanations:

> (…) Natural Sciences, by articulating various fields of knowledge, needs to assure elementary school students' access to the diversity of scientific knowledge built through history, as well as take them gradually closer to scientific processes, practices, and procedures of investigation. (MEC, 2018, p. 319)

Finally, the introductory text stated some consequences of arguing in sciences that do not follow directly. First, according to the text, participating in scientific investigations, using observation, amplifying curiosity, using critical thinking, being agent, standing by their points of view, and looking for one's well-being and health (physical, mental, sexual, and reproductive) would directly teach students what is and how to attain social justice and how to work collaboratively. Second, learning how to argue in science would increase one's socio-environmental consciousness, respectfulness, and acceptance of diversity.

5 Discussion and Implications

We aimed to understand the spaces for argumentation in the Brazilian national curriculum and the possibilities given Brazilian educational policies for teachers to teach argumentation in science education through an interdisciplinary work profiting from all languages (scientific, Portuguese, and English). Once BNCC emphasized integral education and argumentation as a duty of all curricular components, we first believed that Brazilian educational policy prescribed the importance of an interdisciplinary approach to the development of argumentation. Unfortunately, our findings do not support this idea.

Portuguese curriculum holds almost exclusively the categories 'Use of argumentative textual marks', 'Rhetorical', and 'Social rules of arguing'. Some aspects of these categories could and should be explored in the Natural Sciences curriculum in a curriculum aligned with scientific literacy. It is important to learn to weigh the best textual marks to use when choosing how to communicate experimental results to better justify the conclusion. It is crucial to scientific endeavour to benefit from the use of qualifiers to persuade an audience about a scientific conclusion. Curiously, no rhetorical categories were found in Natural Sciences. The persuasion of scientific arguments emerges from the evidence used to support a claim. Finally, it is plausible to think of using socio-scientific issues in Natural Sciences to explain what is a debate and its rules. On the other hand, Natural Sciences focus on using evidence and scientific information to construct and evaluate arguments. This may be due to the way different areas conceive the idea of arguing (e.g. what is an argument and what is argumentation), but may also result from the lack of dialogue between BNCC's authors that, in its turn, result in the disciplinary fragmentation of a skill that is desired to be developed in an interdisciplinary fashion.

The almost complete absence of the category 'Identification of one or more arguments' is troubling since we consider the identification of an argument an important initial step to arguing. We expected to find identification skills in younger years and evaluation skills in the older years in all three languages. Also, we expected to find complementary and superposed skills between the disciplines. Recognizing what counts as an argument and its parts is key to learning how to argue (Kuhn, 1993), and recognizing alternative claims is key to any argumentative dialogue (Simon et al., 2006).

The Natural Sciences curriculum has proportionally more argumentation skills than the others. Planning argumentation teaching may represent a challenge for science teachers if they have to address argumentation skills and address the learning of specific content of Natural Sciences. If our goal is to educate individuals capable of using and evaluating arguments and their parts, teachers who work in the STEM perspective need to keep in mind that the sciences curriculum may not be enough to teach argumentation. Thus, collaborative planning and teaching with language teachers are fundamental actions to allow the shaping of the prescribed curriculum into a coherent curriculum in action.

Teachers' agency is also necessary if we consider that the skills are not evenly nor hierarchically distributed in the curricula analyzed. For example, the skill 'Construction of one or more arguments' is a fairly complex skill (it needs the support of other argumentation skills, such as I, II, III, and IV), it appears in all three curricula, appears in early school years, and is the most frequent skill in the Natural Sciences Curriculum. Because it appears in all three curricula, it may be a point of convergence between these three subjects' teachers and enable its development throughout elementary education. Only the Portuguese curriculum has simpler argumentation skills, so the potential contribution from Portuguese teachers to argumentation teaching must be valued. The lack of knowledge of other disciplines' curricula, the difficulty of working from an interdisciplinary perspective, and the resistance to using student-centered approaches (such as argument-based learning) are still barriers that science teachers must overcome.

These findings have implications for how science teachers may seize elements from language disciplines to accomplish the integral education and interdisciplinarity tenets put forward in Brazilian educational policy. We are aware that curricula do not strictly define teachers' planning for each lesson or activity, but the particular meanings are defined by the curriculum (Hodson, 2014). Activities might be interdisciplinary, irrespective of how the national curriculum portrays the disciplines; yet, curricula could influence, foster, promote, or boost interdisciplinarity, and teaching as argument (Kuhn, 2010) is a grand approach to make it happen. If the dialogue between Portuguese and Natural Sciences is not clear in the curriculum, the complementary effect might not be achieved.

Brazilian's national curriculum has been criticized based on the idea that uniting physics, chemistry, and biology as Natural Sciences in high school education did not make the curriculum immediately interdisciplinary (Aguiar, 2019). Elementary education in Brazil has historically put Natural Sciences together, yet, from a STEM perspective, it is licit to question the difference between calling it Natural Sciences and integrating sciences with other disciplines. Our findings unveil a fragmented curriculum when it comes to argumentation. It appears to be the teachers' job to integrate argumentation in Natural sciences as the same argumentation in Languages. Although BNCC highlights the need for an integral education aiming at students' autonomy and the importance of interdisciplinarity (MEC, 2018), we argue that the document contradicts its pedagogical principles and the broader educational policy. It might be an interesting sequel for science education research to understand in which ways STEM education prompts interdisciplinarity better and how argumentation is portrayed in documents that embrace STEM.

From our findings, we would like to highlight the fact that argumentation in the Natural Sciences curriculum is strongly based on evidence and does not give way to different interpretations for the same evidence or phenomenon. This not only is different from other praised curricula (i.e. National Science Education Standards, NRC, 1996; Next Generation Science Standards, NRC, 2013) document, but it also challenges Nature of Sciences' research that values the importance of weighing evidence goes hand in hand with choosing between plausible explanations (Allchin, 2013).

We insist that it can be left for teachers to take that step in the classroom, but the curriculum could indicate this as a goal and help teachers work it out. Teachers might lose themselves in navigating a new curriculum after each educational reform, especially if they do not receive institutional support during the process. In the past fifteen years, educational reforms have characteristically been imprinted with learning outcomes curricula and the criticism about this type of documents are driven by its perception of what quality in education means, how to assess education and to what extent teachers should be agent and autonomous (Mølstad & Prøitz, 2019), valuing results instead of learning processes. Without clarifying why and how each epistemic practice (e.g. arguing) must be implemented in science education, the curriculum can become oppressing. Teachers, as a result, may not be able to foster learning opportunities aligned with the curriculum (Bismack et al., 2014).

6 Concluding Remarks

As we discussed how languages (scientific, Portuguese, and English) in the Brazilian curriculum could combine forces to improve argumentation teaching, we concluded that even though argumentation is present in the curriculum, it does not make it clear for the teachers how to articulate the different languages for one to teach argumentation in Natural Sciences. Our findings indicate that the Brazilian curriculum does not help teachers seeing the spaces in the curriculum to foster the teaching of argumentation by not showing how Portuguese and Natural Sciences could complement each other, by not stating how interdisciplinarity could be fostered using argumentation, or by not embracing STEM aspects regarding arguments. In other words, the Brazilian newest curricular policy may reinforce old language problems in the country by making it difficult for teachers to tackle argumentation in an interdisciplinary fashion which may, in turn, implicate difficulties for STEM and multilingual contexts.

We see argumentation as a means of reasoning that can foster socio-cultural inclusion, autonomy development, and access to scientific information that surrounds the everyday lives of these children. BNCC fragmented organization without clues on how to integrate disciplines hinders the achievement of its pedagogical principles and Brazilian educational policy. Considering argumentation, BNCC affects negatively the development of states' and districts' curricula, the production of pedagogical resources, teacher education, and the assessment.

We hope that these reflections can contribute to the appropriate and intentional planning argumentation teaching for the development of STEM knowledge and skills in multilingual contexts. Studies like this are relevant to highlight that argumentation can help in the learning of STEM concepts (Mathis et al., 2017). Further studies in other countries' curricula are urgently needed, as future generations will be impacted by educational policies that incorporate more explicit considerations about argumentation and its relevance to STEM education in a multicultural world since migration is a beneficial phenomenon for migrants and destination countries (IOM, 2017).

Acknowledgements The authors are thankful for the scholarship of the second author provided by Coordination for the Improvement of Higher Level Personnel (CAPES) and for the preliminary reviews by Maíra Batistoni e Silva.

References

Aguiar, M. A. S. (2019). Reformas Conservadoras e a "Nova Educação": Orientações Hegemônicas no MEC e no CNE. *Educação & Sociedade, 40,*. https://doi.org/10.1590/es0101-733020192 25329.

Allchin, D. (2013). *Teaching the nature of science: Perspectives and resources.* St. Paul, MN: SHiPS Education Press.

Anderson, G., & Arsenault, N. (2001). *Fundamentals of educational research.* London: Routledge Falmer.

Bismack, A. S., Arias, A. A., Davis, E. A., & Palincsar, A. S. (2014). Connecting curriculum materials and teachers: Elementary science teachers' enactment of a reform-based curricular unit. *Journal of Science Teacher Education, 25*(4), 489–512. https://doi.org/10.1007/s10972-013-9372-x.

Bricker, L. A., & Bell, P. (2008). Conceptualizations of argumentation from science studies and the learning sciences and their implications for the practices of science education. *Science Education, 92,* 473–498. https://doi.org/10.1002/sce.20278.

Brown, B. A., Reveles, J. M., & Kelly, G. J. (2005). Scientific literacy and discursive identity: A theoretical framework for understanding science learning. *Science Education, 89*(5), 779–802. https://doi.org/10.1002/sce.20069.

Cohen, L., Manion, L., & Morrison, K. (2011). *Research methods in education* (6th ed.). New York and London: Routledge.

Dass, P. M. (2015). Teaching STEM effectively with the learning cycle approach. *K-12 STEM Education, 1*(1), 5–12.

Dolina, I. B., & Cecchetto, V. (1998). Facework and rhetorical strategies in intercultural argumentative discourse. *Argumentation, 12,* 127–181.

Driver, R., Newton, P., & Osborne, J. (2000). Establishing the norms of scientific argumentation in classrooms. *Science Education, 84*(3), 287–312. https://doi.org/10.1002/(SICI)1098-237X(200 005)84:3%3c287:AID-SCE1%3e3.0.CO;2-A.

Erduran, S., Ozdem, Y., & Park, J. Y. (2015). Research trends on argumentation in science education: A journal content analysis from 1998–2014. *International Journal of STEM Education, 2,* 5. https://doi.org/10.1186/s40594-015-0020-1.

Ezzy, D. (2002). *Qualitative analysis: Practice and innovation.* London: Routledge.

Fang, Z. (2006). The language demands of science reading in middle school. *International Journal of Science Education, 28*(5), 491–520. https://doi.org/10.1080/09500690500339092.

Gimeno-Sacristán, J. (2017). *O currículo: uma reflexão sobre a prática.* Porto Alegre: Penso.

Hodson, D. (2014). Learning science, learning about science, doing science: Different goals demand different learning methods. *International Journal of Science Education, 36*(15), 2534–2553. https://doi.org/10.1080/09500693.2014.899722.

IOM. (2017). *World migration report 2018.* International Organization for Migration (IOM), The UN Migration Agency. Retrieved from https://publications.iom.int/system/files/pdf/wmr_2018_en. pdf.

Jiménez-Aleixandre, M. P., & Erduran, S. (2008). Argumentation in science education: An overview. In: S. Erduran & M. P. Jiménez-Aleixandre (Eds.), *Argumentation in science education: Perspectives from classroom-based research* (pp. 3–27). Dordrecht: Springer.

Kelly, T. R., & Knowles, J. G. (2016). A conceptual framework for integrated STEM education. *International Journal of STEM Education, 3,* 11. https://doi.org/10.1186/s40594-016-0046-z.

Kelly, G. J., & Licona, P. (2018). Epistemic practices and science education. In M. R. Matthews (Org.), *History, philosophy and science teaching: New perspectives* (pp. 139–165). Cham: Springer International Publishing. https://doi.org/10.1007/978-3-319-62616-1_5.

Kelly, G. J., & Takao, A. (2002). Epistemic levels in argument: An analysis of university oceanography students' use of evidence in writing. *Science Education, 86,* 314–342. https://doi.org/10. 1002/sce.10024.

Kuhn, D. (1991). *The skills of argument.* Cambridge: Cambridge University Press.

Kuhn, D. (1993). Science as argument: Implications for teaching and learning scientific thinking. *Science Education, 77*(3), 319–337. https://doi.org/10.1002/sce.3730770306.

Kuhn, D. (2010). Teaching and learning science as argument. *Science Education, 94*(5), 810–824. https://doi.org/10.1002/sce.20395.

Kuhn, D., Arvidsson, T. S., Lesperance, R., & Corprew, R. (2017). Can engaging in science practices promote deep understanding of them? *Science Education, 101*(2), 232–250. https://doi.org/10. 1002/sce.21263.

Lee, M., Wu, Y., & Tsai, C. (2009). Research trends in science education from 2003 to 2007: A content analysis of publications in selected journals. *International Journal of Science Education, 31*(15), 1999–2020. https://doi.org/10.1080/09500690802314876.

Lin, T. C., Lin, T. J., & Tsai, C. (2014). Research trends in science education from 2008 to 2012: A systematic content analysis of publications in selected journals. *International Journal of Science Education, 31*(8), 1346–1372. https://doi.org/10.1080/09500690802314876.

Martín-Páez, T., Aguilera, D., Perales-Palacios, F. J., & Vílchez-González, J. M. (2019). What are we talking about when we talk about STEM education? A review of literature. *Science Education, 103*(4), 799–822. https://doi.org/10.1002/sce.21522.

Mathis, C. A., Siverling, E. A., Glancy, A. W., & Moore, T. J. (2017). Teachers' incorporation of argumentation to support engineering learning in STEM integration curricula. *Journal of Pre-College Engineering Education Research (J-PEER), 7*(1), 6.

MEC. (2018). *Base Nacional Curricular Comum. Ensino Fundamental.* Brasil, Brasília: Conselho Nacional de Educação; Câmara de Educação Básica.

Mendonça, P. C. C., & Justi, R. S. (2013). Ensino-Aprendizagem de Ciências e Argumentação: Discussões e Questões Atuais. *Revista Brasileira de Pesquisa em Educação em Ciências, 13*(1), 187–216.

Mercier, H. (2011). On the universality of argumentative reasoning. *Journal of Cognition and Culture, 11*(1), 85–113.

Mizell, S., & Brown, S. (2016). The current status of STEM education research 2013–2015. *Journal of STEM Education, 17*(4), 52–56.

Mølstad, C. E., & Prøitz, T. S. (2019). Teacher-chameleons: The glue in the alignment of teacher practices and learning in policy. *Journal of Curriculum Studies, 51*(3), 403–419. https://doi.org/10.1080/00220272.2018.1504120.

National Research Council. 1996. National Science Education Standards. Washington, DC: The National Academies Press. https://doi.org/10.17226/4962.

National Research Council. 2013. Next Generation Science Standards: For States, By States. Washington, DC: The National Academies Press. https://doi.org/10.17226/18290.

OECD, Organisation for Economic Co-operation and Development. (2019). *PISA 2018 results.* Brazil—Country Note. Retrieved from https://www.oecd.org/pisa/publications/PISA2018_CN_BRA.pdf.

Ryu, S., & Sandoval, W. A. (2012). Improvements to elementary children's epistemic understanding from sustained argumentation. *Science Education, 96,* 488–526. https://doi.org/10.1002/sce.21006.

Sadler, T. D. (2004). Informal reasoning regarding socioscientific issues: A critical review of the research. *Journal of Research in Science Teaching, 41*(5), 513–536. https://doi.org/10.1002/tea.20009.

Scarpa, D. L., & Trivelato, S. L. F. (2013). Movimentos entre a cultura escolar e cultura científica: análise de argumentos em diferentes contextos. *Magis. Revista Internacional de Investigación en Educación, 6*(12), 69–85. Retrieved from https://www.redalyc.org/articulo.oa?id=2810/281029756005.

Simon, S., Erduran, S., & Osborne, J. (2006). Learning to teach argumentation: Research and development in the science classroom. *International Journal of Science Education, 28*(2), 235–260. https://doi.org/10.1080/09500690500336957.

Toulmin, S. (1958). *The uses of argument.* Cambridge: Cambridge University Press.

UNESCO. (2013). *Glossary of curriculum terminology.* International Bureau of Education (UNESCO-IBE). Retrieved from http://www.ibe.unesco.org/sites/default/files/resources/ibe-glossary-curriculum.pdf.

Van Eemeren, F. H., Grootendorst, R., & Snoeck Henkemans, A. F. (2002). *Argumentation: Analysis, evaluation, presentation.* Mahwah, NJ: Erlbaum.

Velasco, P. del N. (2009). Sobre a Crítica Toulminiana ao Padrão Analítico-dedutivo de Argumento. *Cognitio, 10*(2), 281–292.

Walton, D. N. (1996). *Argumentation schemes for presumptive reasoning.* Mahwah: Lawrence Erlbaum Associates.

Zohar, A., & Nemet, F. (2002). Fostering students' knowledge and argumentation skills through dilemmas in human genetics. *Journal of Research in Science Teaching, 39*(1), 35–62. https://doi.org/10.1002/tea.10008.

Renata de Paula Orofino is an Assistant Professor at Federal University of ABC, Brazil, engaged in argumentation analysis from a sociocultural perspective and interested in understanding the role of scientific argumentation in promoting social justice, focused on scientific subjects that impact societies' interpretations of gender, race and disabilities in science and in science education.

Nathália Helena Azevedo is a Ph.D. candidate in science teaching at the University of São Paulo, Brazil, and Fulbright alumni. Currently is engaged in science education research, focused mainly in nature of science and curriculum. Works with didactic materials and is interested in projects related to human rights.

Daniela Lopes Scarpa is a Professor at University of São Paulo, Brazil, engaged in science education research, focused on argumentation, nature of science, inquiry-based science teaching and scientific literacy in the context of teacher training and public policy.

Principles for Curriculum Design and Pedagogy in Multilingual Secondary Mathematics Classrooms

William Zahner, Kevin Pelaez, and Ernesto Daniel Calleros

Abstract We introduce and illustrate three principles for designing secondary mathematics classrooms in which multilingual students can benefit from participating in mathematical discussions. Drawing from the Academic Literacy in Mathematics (ALM) framework (Moschkovich, 2015), we developed these principles through a four-year design research collaboration with ninth grade mathematics teachers working in a linguistically diverse urban secondary school in the southwest USA. The three principles that we developed through this work are (a) Align the conceptual focus and use of problem contexts across each curricular unit, (b) Integrate practice-focused and content-focused learning goals in a trajectory, and (c) Incorporate structures that enable the widest possible participation in classroom discourse. We elucidate the necessity for and implementation of each principle by presenting illustrative cases from our research. At the conclusion of the chapter, we consider the implications of this work for STEM policy and practice in multilingual settings.

Keywords Academic Literacy in Mathematics · Secondary mathematics · Design research

1 Introduction

Participating in classroom discussions can promote student learning in multilingual STEM classrooms (e.g., O'Connor et al., 2015). Through engaging in discussions, emergent multilingual students, including students who are in the process of learning the language of schooling, can simultaneously develop disciplinary understandings and appropriate STEM discourses and practices (Moschkovich, 2015). However, teachers in linguistically diverse STEM classrooms must plan carefully to ensure that emergent multilingual students can fully participate in classroom discussions (Adler, 2001; Adler & Ronda, 2015; Barwell et al., 2017; Chval et al., 2014; Setati & Adler, 2000). In this chapter, we introduce and illustrate three principles we have

W. Zahner (✉) · K. Pelaez · E. D. Calleros
San Diego State University, San Diego, CA, USA
e-mail: bzahner@sdsu.edu

© The Author(s), under exclusive license to Springer Nature Switzerland AG 2021
A. A. Essien and A. Msimanga (eds.), *Multilingual Education Yearbook 2021*, Multilingual Education Yearbook,
https://doi.org/10.1007/978-3-030-72009-4_13

developed for designing secondary mathematics classrooms in which multilingual students can benefit from classroom discussions:

1. Align the conceptual focus and use of problem contexts across each curricular unit.
2. Integrate practice-focused and content-focused learning goals in a trajectory.
3. Incorporate structures that enable the widest possible participation in classroom discourse.

These principles were developed through a design research effort situated in a linguistically diverse urban secondary school in the Southwest US. The goal of the study was to redesign the ninth grade mathematics classroom to create a learning environment in which emergent multilingual students could participate in classroom discussions. The four-year research effort was a collaborative effort of ninth grade mathematics teachers, researchers, and graduate student researchers. It is our contention that both the methods of this work and the principles abstracted from our design research can inform educators and policymakers who design STEM learning environments for multilingual students.

In what follows, we introduce the theoretical framework for this design and development effort, and we connect the framework to our design principles. Then, we present the study and the research context. Next, we describe and illustrate the three design principles. For each principle, we present examples from our research to illustrate what the principle looked like in practice. At the conclusion of the chapter, we consider the implications for policy and practice in multilingual STEM learning environments.

2   Theoretical Perspective and Design Framework

Our overarching theoretical framework is grounded in a sociocultural approach to learning, where learning is conceptualized as developing participation in culturally shared practices (Forman, 1996; Moschkovich, 2002). The sociocultural theoretical framework highlights that language(s) and discourse practices are both mediators of, and targets for learning (Forman, 1996). In alignment with this theoretical stance, we drew upon the Academic Literacy in Mathematics (ALM) Framework (Moschkovich, 2015) to frame our design principles. Within the ALM framework, are three interrelated dimensions: (a) mathematical proficiencies, (b) mathematical practices, and (c) mathematical discourses (Moschkovich, 2015).

Mathematical proficiencies are the forms of expertise, knowledge, and skill that are developed in school mathematics (National Research Council, 2001). Of particular interest to us is the interplay between *procedural fluency* and *conceptual understanding*. Due to restrictive language policies and patterns of tracking, many emergent multilingual students in the US are placed in low-track mathematics classes where learning focuses on procedures (Callahan, 2005; Kanno & Kangas, 2014). *Mathematical practices* are culturally organized activities shared by members of the mathematical community (Lave & Wenger, 1990). In mathematics, disciplinary practices include problem solving, argumentation, and modeling. *Mathematical discourse* is the ways people use language(s) and semiotic systems to do mathematics. The term discourse indicates that learning "the language of mathematics" entails far more than acquiring mathematical vocabulary or even learning the mathematics register (Moschkovich, 2002). In particular, learning and participating in the Discourse[1] practices of a community (Gee, 1996) are intertwined with developing a disciplinary identity.

We focused our design efforts on promoting classroom discussions because there is alignment between the goals of fostering mathematical discussions and developing ALM. According to Chapin et al. (2014), the goals of classroom mathematical discussions are helping students (a) clarify their mathematical thinking, (b) orient to the thinking of others, (c) deepen their thinking, and (d) engage with the thinking of others. These purposes align with the dimensions of ALM. For example, clarifying and deepening one's thinking align with the ALM dimensions of developing *proficiency* and engaging in mathematical *practices*. Orienting and engaging with the thinking of others are aligned with the ALM dimensions of developing mathematical *practices* and mathematical *discourse.*

Moschkovich and Zahner (2018) show how mathematical discussions in multilingual settings could be analyzed through the lens of the ALM framework. They also suggested the ALM framework can be used to design lessons for linguistically diverse classes. In this project, we grounded our design principles using the dimensions of ALM as an organizing framework. We note, however, that the dimensions of the ALM framework are closely related and interdependent. Given this interdependence, some dimensions of the ALM apply to more than one design principle, and vice versa. Table 1 contains a summary of the design principles, a rationale for their use, and connection to the ALM framework.

3   Methodological Framework

Our methodological framework arises from design research (DR; Design-Based Research Collective, 2003). DR is an iterative and interventionist methodology where researchers purposefully design a learning environment, explore phenomena that emerge as a result of the design, and refine the design for future iterations (Prediger et al., 2015). For this project, design research was appropriate because disciplinary discussions in linguistically diverse classrooms are relatively rare (Chval et al., 2014; Zahner et al., 2012; Zahner, 2015), and unlikely to arise without intentional effort and intervention (Setati & Adler, 2000; Prediger et al., 2015).

---

[1]Gee uses a capital D to denote the distinction between discourse as a unit of text and a Discourse as a socially recognized way of using language to signal membership in a community of practice.

Table 1 Design principles

Design principle	Rationale	Primary correspondence with ALM
(1) Align the conceptual focus across the curriculum and carefully choose problem contexts	Reduce unnecessary linguistic demand and allow students to build experiences to reason mathematically (Chval et al., 2014). We elected to use one conceptual focus (slope as rate of change) and one recurring context (running races) across the unit	Development of mathematical proficiencies, particularly a conceptual understanding of slope as rate of change
(2) Integrate practice-focused and content-focused learning goals in a trajectory	Specify goals that are linked to mathematical practices and specific to the content (Prediger & Zindel, 2017). While we focused on engagement in practices, we called these *"language goals"* to align with the terminology in the school's lesson planning documents	Engagement in mathematical practices, particularly explaining, justifying, and using/relating representations
(3) Incorporate structures that enable the widest possible participation in classroom discourse	Engage a wide spectrum of students, including English learners, in classroom discussions by creating structures to facilitate classroom talk. Structures included Mathematical Language Routines (Zwiers et al., 2017), the use of dynamic representational technology (Zahner et al., 2012), and teacher talk moves (Chapin et al., 2014)	Participation in mathematical discourse, especially in verbal and written modes of expression

To narrow the scope of our design effort, we focused on a single unit of instruction, the introduction of slope in the ninth grade mathematics curriculum. This focus was chosen because slope is a critical topic that spans elementary mathematics through calculus (Thompson, 1994), it is used across STEM disciplines, and because communicating about slopes requires semiotic coordination of quantities (Lobato & Ellis, 2010).

This design effort was intentionally situated in a "typical" school attended by multilingual learners in the US. We also attempted to fit our designs within the daily constraints faced by teachers. Thus, the pace of our design cycles was bound by the school's mathematics curriculum and yearly content pacing guides. Therefore, the iterations of design research occurred across multiple academic years. Figure 1 shows the project timeline, which included three phases of classroom observations

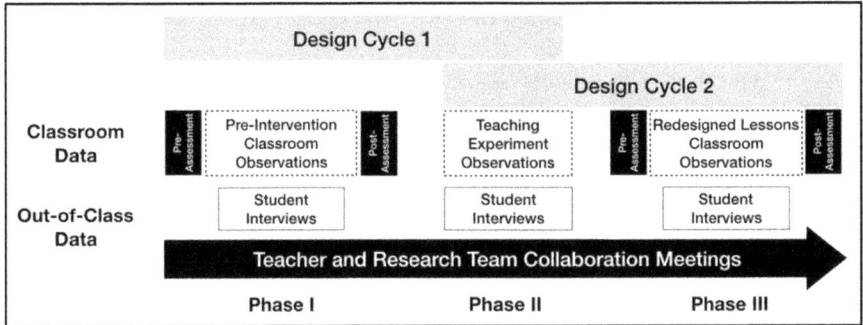

Fig. 1 Phases of the design experiment

across three academic years, and two iterative design cycles. Each set of classroom observations lasted about nine class meetings, coinciding with the introduction of slope.

Additionally, one unique feature of this project was the involvement of the participating teachers in the research and design effort. Our commitment to equity was reflected in our choice to work *with* the teachers at the research site, rather than to design lessons *for* the teachers. The teachers were an integral part of the design process from start to finish. As the project concludes, the teachers are contributing to the dissemination efforts (e.g., Zahner et al., 2018), and are sharing the work in their local school system.

4 Context, Data, and Analysis

4.1 Setting

The setting of this research effort was three ninth grade integrated mathematics classes at City High (a pseudonym), a large, linguistically diverse, urban high school in the US–Mexico border region. City High serves a student population in which nearly all students are from non-dominant racial/ethnic groups, and the vast majority of students are from working class and poor families. At least 80% of all students at City High were multilingual learners: about 30% were classified as English Learners (ELs), and 50% of students were formerly classified as ELs at some point in their K-12 education. We note that the label EL focuses on what students *lack* rather than their *resources*. In this paper, we use multilingual students and emergent multilingual students to highlight students' assets, and we use EL when describing students' assigned language classification. Spanish was the most common primary language among City High students. In terms of race and ethnicity, 77% of students were identified as Latinx, 12% Asian, 7% African American, and 4% other. About 89%

of students were from low-income families. The classes within which we worked reflected these demographics.

As is typical in US schools serving minoritized and low-income students, City High had high rates of student absence and mobility, high teacher turnover, and low average scores on state-mandated standardized assessments. As teachers committed to promoting equity, we chose to do this research in a setting reflective of the typical schools that the vast majority of US emergent multilinguals attend (Gándara & Contreras, 2009). The research team intentionally included bilingual researchers who were from the local community and who were familiar with the sociocultural context of the school, neighborhood, and city.

4.2 Data Sources

The primary data sources for the teaching events were video recorded ethnographic classroom observations. The field observers recorded fieldnotes and collected visual aids used by the teachers. Additional data sources (see Table 2) included written assessments taken by the students at the start and conclusion of each unit, and interviews with students from each class.

In separate works, we present findings from the analyses of student learning as measured by the pre-post written assessments (Zahner et al., 2020). We are also preparing works on student perspectives expressed during the interviews and in-depth qualitative analysis of classroom data (Zahner et al., 2021). In order to be concise, we do not discuss the assessments or student interviews further in this chapter.

Table 2 Data sources and analyses

Data source	Phase	Forms of data collected	Analysis
Pre-post assessments	1, 3	Written assessment	Quantitative analysis of pre-post-gains by question type (procedural or conceptual) and student language (EL or non-EL)
Student interviews	1, 3	Digital videos and transcripts; student written work	Qualitative analyses of mathematical reasoning and student identity
Classroom observations	1, 2, 3	Field notes; digital videos and transcripts; images of student written work	Classroom discourse analysis; examination mathematical discourse practices

4.3 Analysis

The analysis of data from the classroom observations was rooted in ethnographic discourse analysis (Gee & Green, 1998), paralleling the methods outlined in Moschkovich and Zahner (2018). As a first step in this process, each classroom observation was summarized in a structured memo that synthesized how the teachers and students participated in discussions (e.g., what problems were discussed, who participated, what representations or resources were used, which ideas were taken up). These summaries allowed us to document how emergent multilingual students used their linguistic resources to reason about mathematics, as well as how the teachers used languages (in the sense of both named languages and discourses) and mathematical semiotic resources to communicate. The ethnographic analysis of the joint activity of the classroom discussion was then carried through by systematically summarizing the material, activity, semiotic, and sociocultural aspects of the situation (Gee & Green, 1998). In line with the principles of design research, the analysis and the development proceeded in an iterative process, and the lesson designs were refined at each stage. Through the inductive research process, we developed the three design principles described in Table 1. In the remainder of this chapter, we elaborate on the definition of each design principle, describe the rationale for its development, and illustrate the principle in our work.

5 Illustrations of the Design Principles

The three principles outlined in this project were intended to support the design of instructional units. By a unit, we mean a collection of lessons focused on one major topic. A unit is more than a single lesson, and considerably less than an academic term or school year. The three principles that we describe here are not intended to be exhaustive. Rather, these design principles are the three most prominent principles that arise in our collaborative work in the local context of City High.

5.1 Principle 1: Align the Conceptual Focus and Use of Problem Contexts Across Each Curricular Unit

We developed Principle 1 based on our desire to develop mathematical proficiencies, particularly conceptual understanding of important mathematical ideas. We identified a coherent conceptual focus, and embedded that focus in lessons using a "realistic" and rich problem context. By *conceptual focus*, we refer to the core meanings, representations, and solution strategies embedded within the unit. The same mathematical topic can be taught with different conceptual foci. For example, the school-adopted curriculum materials at City High introduced three distinct meanings for slope: steepness, rate of change, and (constant) slope as a defining feature of linear functions. In accordance with Principle 1, we selected one of these meanings as the conceptual focus.

A *problem context* refers to the story or situation within which a mathematical problem is posed. Some problems in school mathematics are set in a purely mathematical contexts while others are "applied" problems. Problem contexts are important for language learners because problems set in unfamiliar contexts can present unnecessary barriers for emergent multilingual students (Martiniello, 2008). Conversely, a well-chosen problem context may be a resource (Chval et al., 2014). For our designed unit, we sought to embed the mathematics in a realistic and rich context that would support language learners' mathematical reasoning. We use *realistic* in the sense of Realistic Mathematics Education (Gravemeijer & Doorman, 1999), meaning a context that students can readily imagine and relate to. By a *rich context*, we mean one that allows students to see multiple mathematical insights related to a phenomenon.

How Design Principle 1 Was Developed

During Phase I of the project, we conducted a conceptual analysis (Thompson, 2008) and reviewed prior research on student learning of slope (e.g., Lobato & Ellis, 2010; Stump, 2001). We also examined the presentation of the topic in the school-adopted curriculum materials. We noted the use of problem contexts by cataloging the contexts that were used in prior research, in the school textbook, and in the teachers' lessons. From this analysis, we found that slope was introduced over four main lessons: (a) introducing slope using the image of "steepness", (b) calculating the slope given coordinate points in a purely mathematical context, (c) interpreting slope as a rate of change in graphs relating quantities from "real life" contexts, and (d) using the slope calculation to check whether sets of points were collinear. We noted that there were at least three different conceptual foci in the unit introducing slope (steepness, rate of change, and defining property of a line). We also noted that between the expository text and exercises, the written curriculum included more than 16 real life and mathematical problem contexts to introduce the topic of slope in four lessons.

How Design Principle 1 Was Realized in the Redesign

After Phase I, we chose to redesign the unit by focusing on the "rate of change" meaning of slope. This choice was informed by our review of the research where this meaning features prominently (Lobato & Ellis, 2010). The rate of change meaning is conceptually deep and the other meanings (e.g., measure of steepness, Stump, 2001) can be derived from the "rate of change" meaning for slope. To develop this meaning, we also chose to focus on one main problem context: reasoning about movement and speed in situations involving motion. This context focus allowed us to (a) develop technology-enhanced tools to support the context-embedded learning (e.g., Thompson, 1994; Zahner et al., 2012) and (b) create a sequence of lessons where the concepts were developed in a trajectory informed by prior research (e.g., Lobato & Ellis, 2010). This choice of context was also informed by our consideration of the sociocultural resources that the multilingual and emergent multilingual students bring to the classroom, and the level of linguistic demand in the curriculum materials. In light of these considerations, reasoning about problems set in the context movement and speed seemed most likely to support the development of multilingual students' mathematical proficiency, particularly their conceptual understanding.

The redesigned unit had nine core lessons. These lessons built on a sequence of lessons aligned with Lobato and Ellis's (2010) trajectory of essential understandings, starting from ratio reasoning, leading to slope as a rate of change, and ultimately leading to writing linear equations. In the first lesson, students experienced a race and identified the quantities that determine "how fast" someone is running. The purpose of this lesson was to focus students' attention on the quantities involved in their reasoning. Next, three lessons focused on introducing representations and identifying different combinations of distance and time that result in "same speed" movement (e.g., 60 meters in 10 s and 30 meters in 5 s are the same speed). These problems were sequenced to introduce and connect multiple representations. The problems were also sequenced to increase in complexity and to necessitate making generalizations. The next two activities required students to apply the concept of rate of change to solve problems in two different contexts. Next, three lessons focused on generalizing the use of linear equations to model motion, including piecewise linear functions presented in an animation. Finally, the students reasoned about and solved problems about average rate of change, and wrote linear functions when initial values were not zero. This sequence of lessons was supported with integrated discourse goals, and intentionally structured interactional routines (described below).

5.2 Principle 2: Integrate Practice-Focused and Content-Focused Learning Goals in a Trajectory

We developed Principle 2 in order to ensure that multilingual students and emergent multilingual students had access to the full range of mathematics learning opportunities that are envisioned in the ALM framework (Moschkovich, 2015). In particular, Moschkovich and Zahner (2018) suggest that planning for mathematics instruction should include both cognitive aspects of mathematical activity (e.g., mathematical reasoning and thinking) as well as sociocultural aspects (e.g., participating in and appropriating mathematical practices and discourses). Such intentional planning is vital in multilingual classrooms since the cognitive and sociocultural aspects develop interdependently. This broad focus on academic literacy extends prior research where a more narrow focus on "academic language" has been identified as the primary obstacle to learning for both monolingual and multilingual students (Moschkovich, 2015). In particular, engaging in mathematical practices such as explanation and justification is "not only a learning goal but also an important design principle for achieving the goal" (Prediger & Zindel, 2017, p. 4167). Without including practice-focused goals (our Principle 2) and discourse support structures (Principle 3, below) designers may unintentionally replicate the existing inequities that privilege students who have access to the "normative" practices and discipline-aligned discourses outside school (Gee, 1996; Prediger & Zindel, 2017).

How Design Principle 2 Was Developed

Our second design principle is intended to guide teachers to consider what opportunities students have to participate in and appropriate mathematical practices. Therefore, each lesson included *mathematical content goals* and *mathematical language goals* (which we considered more broadly to be *practice* goals, as explained below). *Mathematical content goals* are topic-specific objectives related to the cognitive aspects of mathematical activity. For example, the mathematical content goal for a lesson in this study was to "identify the average rate of change using a piecewise linear graph." By itself, the mathematical content goal does not reflect the mathematical practices that may be present in authentic mathematical activities when identifying an average rate of change (Moschkovich, 2015). If the content goal is considered alone, a teacher may possibly overemphasize the use of a procedure or algorithm to calculate slopes (Zahner, 2015).

Yet, the intertwined dimensions of the ALM framework point to a desire for students to engage in disciplinary practices while engaging in the cognitive aspects of mathematical activity. Therefore, in each lesson we developed, we also included *mathematical language goals*. These were goals that highlighted targets for developing disciplinary practices related to specific mathematical content goals (Prediger & Zindel, 2017). In the case of average rates of change, the language (practice) goals were to "make generalizations" about calculating average rates and to "describe processes." We called these practice-focused goals language goals since the teachers at City High were required to include "language goals" in their school lesson plans. But, even with this name, our focus was broader than developing language in isolation. We did not use language goals such as "define and use the word rate in a sentence." Rather, we developed language goals focused on using academic language to engage in valued mathematical practices such as explaining, justifying, or representing. These practice-focused goals are language goals in the sense that engagement in practices requires students to use mathematical language and symbols.

How Design Principle 2 Was Realized in the Redesign

When designing each lesson, we collaborated with the teachers to identify one or two valued mathematical practices related to the core mathematical focus of the lesson. Revisiting the example above, in a lesson on average rates of change, one of the language goals was for students to "make generalizations about multiplicative distance-time-average speed relationship in piecewise linear graphs." This goal includes the practice of explaining, and the explanation is related to the content—identifying the average rate of change. The intention was to prompt students to use written language and symbolic resources (e.g., a slope triangle drawn on a graph) to make and support their generalizations. One ideal student response might include an explanation with a graph: "I can find the average rate by making a triangle using the two endpoints to find distance and time, then we divide the y difference, or distance, by the x difference, time."

When writing the language/practice goals in alignment with the ALM framework (Moschkovich, 2015), we considered mathematical language goals with different language functions across the unit. *Language functions* include verbs that describe the intended mathematical practices to be developed in class activities, such as asking students to EXPLAIN the process about calculating the average rate and GENER-ALIZE about relationships. At times, teachers interpret language goals as implying that they must stop teaching mathematics and teach a language lesson in isolation. To circumvent this, we also considered ways to expand language goals to include multiple practices and to encourage teachers to allow students to use different modes of communication (oral, written, receptive, and productive), symbol systems (e.g., written text, numbers, graphs, and tables), as well as the language(s) of their choice while engaging in these practices (Moschkovich, 2015).

Finally, in this section we must note that there are some mathematical practices that are non-verbal and not necessarily captured by a "language goal." For example, a teacher may want their students to develop the practice of *perseverance*. Thus, we acknowledge that our strategic decision to use the term "language goals" (which was made to align with the constraints of City High) may be unnecessarily limiting. In future iterations of this work, we will likely rename Principle 2 to highlight the centrality of practices as our target for development.

5.3   Principle 3: Incorporate Supportive Language Structures to Enable the Widest Possible Participation

Our final design principle entailed providing supportive language structures to foster the simultaneous development of mathematics and language among a broad group of students in our multilingual setting. The primary structures we used included inter-active technology, explicit language supports, and the repeated use of interactional routines to promote student participation in mathematical discourse. Principle 3 is primarily aligned with the mathematical discourse dimension of the ALM frame-work. Our goal for including this principle in our designs was to make participation in mathematical discourse available to a wide group of students in our multilingual setting.

Principle 3 is necessary because, to realize the mathematical language goals in our lessons, students must have access to mathematical discourse, "the communicative competence necessary and sufficient for competent participation in mathematical practices" (Moschkovich, 2015, p. 47). As Moschkovich (2015) noted, this commu-nicative competence involves successfully navigating and relating multiple semiotic systems (natural language, mathematics symbol systems, and visual displays), as well as being able to use language(s) in different ways and for different purposes, including using multiple modes, multiple representations, different types of written texts, different types of talk, and addressing different audiences. To meet these broad requirements for developing communicative competence among emergent multilin-gual students, it is vital to provide structures that allow students, including those learning the language of instruction, to coordinate multiple semiotic systems and participate in different forms and settings.

In turn, allowing for diverse ways of participating and with different audiences invites a larger diversity of students to engage in mathematical discourse. This is especially important in linguistically diverse classrooms, where different students possess varied communicative resources, which they may apply differently depending on the talk environment and audience surrounding them (Chval et al., 2014). For example, a multilingual student may feel more comfortable using their primary language when they communicate one-on-one with another multilingual student than when they communicate in a discussion involving the whole class.

How Design Principle 3 Was Developed

Our Phase I observations revealed that the mathematics learning environments at City High provided students with few communicative opportunities, and little change in modes of communication or audiences. For example, one teacher devoted the overwhelming majority of class time to whole-class lecturing. The teacher presented the content and the students were expected to learn by watching and listening to the teacher, and occasionally answering short questions in verbal form. There were no opportunities for students to talk to their peers, communicate in writing, or use multiple representations in the service of communicating. These rigid communicative environments limited student opportunity to develop mathematical discourse practices, potentially discouraging students' identity development as doers of mathematics (Boaler & Greeno, 2000).

Hence Principle 3 centers on incorporating supportive language structures that allow students to coordinate multiple semiotic systems, participate in multiple ways, and engage in mathematical practices. Three language structures that we designed into the Phase II and III lessons included: (a) alternating whole class and small group formats (Zahner et al., 2012), (b) using "mathematical language routines" (Zwiers et al., 2017), and (c) incorporating technology and dynamic representations (Zahner et al., 2012).

How Design Principle 3 Was Realized in the Redesign

Alternation of Whole Class and Small Group Discussions. Each lesson plan in the unit included intentional alternation of discussion formats. Whole-class discussions are those in which a teacher or another presenter (e.g., a student making a presentation) is an active facilitator. In contrast, for small-group discussions, the teacher assigns students into small groups of two to four students and gives them a mathematical task to talk about within their group. Both talk formats have advantages and disadvantages that make them more effective when they are used in an alternating fashion. Whole-class discussions allow for a large variety of ideas to be explored. However, whole-class discussions may also be too large and public to provide some students, especially multilingual students who are developing fluency in the "official" language of school, with the more low-stakes, collaborative, and personalized space that small group discussions may afford. Small-group discussions are especially useful for multilingual students because they are more private, which may allow students to use resources, such as their primary or preferred language(s), that they might not feel as comfortable sharing during whole-class discussions (Setati &

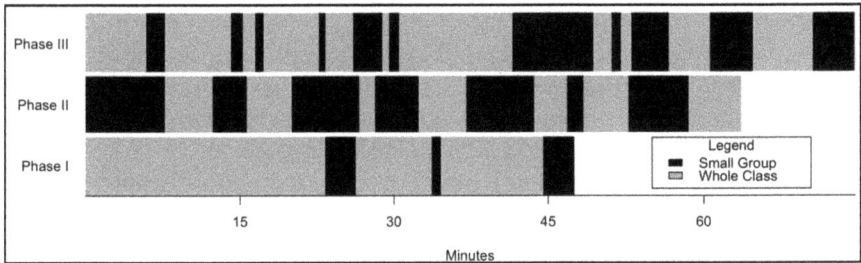

Fig. 2 Timelines showing the lesson format during three lessons

Adler, 2000; Zahner, 2012). Used in conjunction, the two talk formats give students access to two aspects of mathematical discourse from ALM (Moschkovich, 2015): two different audiences (the teacher and all peers in the class, and a small group of peers) and two different types of talk (exploratory and expository). In our designs, we intentionally alternated between whole class and small group format in the lesson structure. Figure 2 shows a timeline of the alternation of whole-class and small group format in one sample class on average rates of change taught by the same teacher across all three phases of the project.

Mathematical Language Routines. To give students access to mathematical discourse, we incorporated *mathematical language routines* (MLRs) in our lesson designs across the unit. MLRs are adaptable language structures to amplify, assess, and develop students' language (Zwiers et al., 2017). These routines are designed to capitalize on and further develop students' agency in their own mathematical and linguistic sense-making. The MLRs used in this study include eight in Zwiers et al. (2017) as well as routines from other researchers (e.g., Driscoll et al., 2016). Below we illustrate three routines: Discussion Supports, Stronger and Clearer Each Time, and Collect and Display.

Discussion Supports are intentional talk moves used to help individual students share their ideas, help students to orient to others' thinking to deepen their own reasoning, and help students build on others' language and ideas (Chapin et al., 2014; Zwiers et al., 2017). Talk moves like revoicing and pressing for reasoning fall into this category of MLR. Discussion Supports also invite more student participation and help students make sense of complex language and ideas involved in mathematical activity by giving students access to various aspects of the mathematical discourse element of the ALM framework (Moschkovich, 2015). After a teacher uses these discussion supports consistently, students begin to take up the discourse practices embodied in the support.

For example, in one of the Phase III lessons, the students were solving a challenge problem. They used a simulation that provided feedback on the correctness of their answers. One group in the class found an answer and a member of the group shared his answer with other students. José and Angel, two multilingual students checked the answer that they were given by their classmate. However, José was not satisfied with just knowing the answer: "Pues dicen que sí, pero tenemos que saber como lo agarraron. [*Well, they say yes [the feedback on the software], but we have to know how they [the other group of students] got it.*]." Jose's statement anticipates the expectation that they must not only produce the answer, but also engage in disciplinary practices such as explaining or justifying that answer. This anticipation was likely a result of the teacher's consistent use of the talk move "press for reasoning."

Stronger and Clearer Each Time is a mathematical language routine that provides an organized opportunity for students to revise their verbal and written ideas through peer interaction (Zwiers et al., 2017). In this routine, students may begin by thinking or writing a response to a mathematical prompt individually. Then, students have opportunities to revise and refine their response by engaging in a structured pairing activity through which they discuss their ideas. Finally, the students revisit their initial written response and revise their work. An example of Stronger and Clearer appeared in a worksheet we used in a lesson about rate of change (Fig. 3). The lesson revolved around comparing rates of change in a graph with four linear functions. In the worksheet, Question 4 was included twice. First students were instructed to write their individual and initial thoughts. Then, students were invited to write a revised and refined version of their explanation after engaging in a structured peer discussion (The icon on the worksheet was a design element to signal the MLR).

The Stronger and Clearer Each Time routine gives students access to at least two aspects of mathematical discourse from the ALM framework: multiple modes of communication (oral, written, receptive, and expressive) and different types of communication (exploratory and expository). Although their official output is in written form, students engage in talk to revise and refine their written output. In turn, to engage productively in talk, students must both receive others' ideas and express

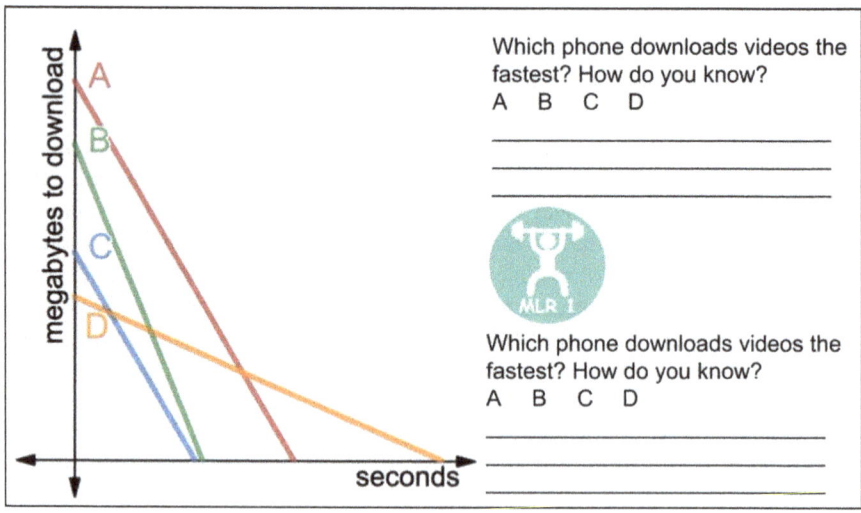

Fig. 3 A segment of a worksheet used to incorporate a stronger and clearer each time mathematical language routine

their own ideas. Their initial and individual thoughts and subsequent talk may be more exploratory in nature, whereas their finalized written output may be more expository. Moreover, through participating in this routine, students have opportunities to engage in valued mathematical practices, such as constructing and critiquing arguments and attending to precision in their communication and ideas.

Collect and Display is a routine whereby the teacher captures students' ideas and language and showcases them by writing, drawing, or presenting them to the entire class (Zwiers et al., 2017). In a Collect and Display routine the focus is not on capturing correct answers, but rather publicly validating student-generated language. With this structure, teachers can help students to connect natural language with visual displays and mathematics symbol systems, thereby helping them relate three semiotic systems involved in mathematical discourse (Moschkovich, 2015). For example, in our design effort, the students were shown an animation of two characters with a linked graph (see Fig. 4). Emma moves at a non-constant rate, while Average Emma moves at a constant rate. Students were asked to play the animation and then individually answer the questions "What do you notice? What do you wonder?" Students could move the points in the graph while exploring this activity, but the relationship between Emma and Average Emma was maintained. Next, the teacher facilitated a whole-class discussion around the questions "What do you notice? What do you wonder?" As the students engaged in discussion, the teacher wrote their ideas on the board (see Fig. 5).

Some language the teacher recorded during this routine included "Average Emma doesn't stop" and that Average Emma was moving at a "constant rate." When the teacher asked other students to build on or restate these ideas, students responded that "the speed doesn't change" and that the speed is "consistent." In this way, the

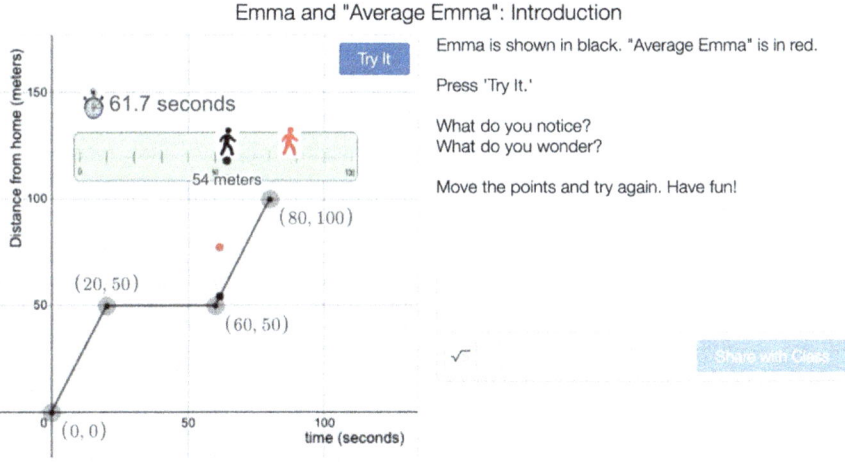

Fig. 4 Mathematical activity preceding a Collect and Display routine in our design effort

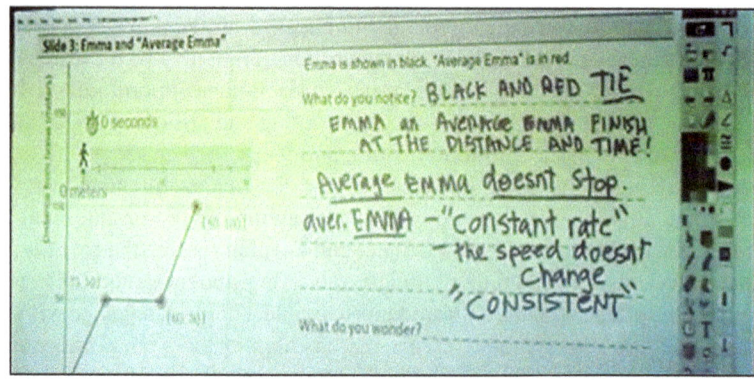

Fig. 5 Teacher's use of a Collect and Display in our design effort

Collect and Display routine allowed students to generate language, and this turned out to be exactly the language that was the goal of the lesson!

This instantiation of Collect and Display illustrates how the language routines are intertwined with each other, and how the routines relate to the discourse dimension of the ALM framework. This routine helped students connect multiple representations (the animation and graph) using everyday and disciplinary language (e.g., "speed doesn't change" and "constant rate"). In terms of the lesson goal, this routine prepared for considering how to find Average Emma's rate, and how that related to Emma's motion. This routine also connects to disciplinary practices in ALM, such as making sense of the problems and modeling with mathematics.

Technology. Finally, we discuss our strategic use of dynamic representations. While there are many useful ways to incorporate technology in a linguistically diverse classroom (e.g., showing videos or using calculators), one particularly important use of technology is in conjunction with dynamic representations (Roschelle et al., 2000; Zahner et al., 2012). This type of dynamic representational technology includes two main features: (a) it displays multiple linked representations, and (b) the representations are editable. In some cases, this kind of technology may also incorporate animations or feedback. The dynamic representational technology we used is shown in Figs. 4 and 5 (interested readers can access one activity here: https://teacher. desmos.com/activitybuilder/custom/5dc341a34194856bb9106d77). Using dynamic technology in the context of discussions prompts can help emergent multilingual students connect mathematics symbols systems and displays with natural language, gestures, actions, and disciplinary terminology (Ng, 2016). Thus, this technology structure can be useful for developing the discourse dimension of ALM by giving students access to complex aspects of mathematical discourse, including the interaction between multiple semiotic systems.

6   Discussion and Conclusion

In this chapter, we have described and illustrated three principles for designing secondary mathematics classroom learning environments to promote the engagement of multilingual students in mathematical discussions. These design principles are grounded in the dimensions of the Academic Literacy in Mathematics framework (Moschkovich, 2015). We developed, tested, and refined these design principles through cycles of design research situated in an urban school serving a linguistically diverse student population. We note that the three design principles presented in this chapter are not intended to be an exhaustive list of all of the forms of scaffolding that might support emergent multilingual students in STEM classrooms. Instead, these principles emerged as vital within the context of our work, and the enactments of these principles were tailored to fit the constraints of our research site.

Revisiting the theoretical and conceptual frameworks that undergird this effort, at times in our team meetings, we struggled to distinguish the principles and design features targeting mathematical practices and mathematical discourse. This is not unexpected: Moschkovich (2015) notes that the dimensions of the ALM framework are interrelated and interdependent. Nonetheless, we found that for the purpose of this project, it was helpful to try to distinguish between our mathematical practice goals (which we called language goals) and the specific discourse supports that we used to meet our goals (e.g., MLRs, alternating whole-class and small group discussions, technology) that were incorporated in the designs of each lesson.

6.1   Implications for Practice

A curious reader may question whether the design principles were effective. That is, did the designs result in better lessons, different forms of participation, or more learning? Space constraints limit our ability to share extensive empirical evidence of the effects of our designs in this chapter. However, in other works (Zahner et al., 2020, 2021) we use both quantitative and qualitative analyses to show some effects of the redesign effort. For example, in Zahner et al. (2020) we analyzed the pre- and post-assessments at each phase of the project, disaggregating the data by student language classification. In that analysis we found that across the Phase III unit, all students made gains on the curriculum aligned assessments from the pre-unit test to post-unit test. But, the largest gains on test questions with a conceptual focus were among students who were classified as ELs. Through detailed qualitative analysis of pivotal moments, we also noted transformations in the patterns of classroom discourse (Zahner et al., 2021) from Phase I to Phase III. However, we also noted that the constraints of the school and classroom settings appeared to shape the implementation of our designs and these constraints may have limited student agency and choice in the redesigned lessons.

Finally, an important aspect of and motivation for this work was our focus on promoting more equitable learning experiences through our design efforts. This equity orientation necessitated working with teachers to co-design (rather than giving teachers lessons). In extending this work, we are considering how we can (a) further challenge the assumptions about the school and classroom environments, and (b) also incorporate the perspectives of multilingual *students* as collaborators in future design work. It is exciting and intriguing to imagine how incorporating the voice and perspectives of students might also contribute to the development of more equitable learning environments.

6.2   Implications for Policy

We note that the design principles and the materials developed within this project were tailored to the context and the content area-focus. The context included the particular sociocultural and sociopolitical setting of a large urban high school located in the US–Mexico border region. This context shaped the classroom interactions we observed in Phase I, and the lesson and unit designs that were developed in response. Thus, we cannot claim that the materials we created in this project will necessarily "work" in other contexts. For us, engaging in the co-design process was a critical part of the project.

While we do not claim that our designs will transfer across settings, given the theoretical grounding of this work, this effort can provide a starting point and useful design principles for design and development work in other multilingual settings and in other STEM content areas. While the specific instantiation of the design principles outlined here will likely look different in other settings and in other STEM content areas, the theoretical basis for our design decisions will likely remain the same. That is, other researchers who are seeking to replicate this work would do well to develop or adapt design principles that address each of the dimensions of the ALM framework. For policymakers and educational leaders, one critical implication of this work is that they should attend to how curricular materials and educational resources afford opportunities for multilingual students to develop disciplinary proficiencies, practices, and discourse practices in STEM classrooms.

Acknowledgements We acknowledge the assistance of research assistants Yessika Gamala, Josue Gonzalez, Brenda Melendrez, Hayley Milbourne, Alicia Prieto, Irania Rivera, Antonio Villarreal, Lynda Wynn, and April Zuniga, and the collaboration of the teachers who are participating in this study.

This research was based upon work supported in part by the National Science Foundation Grant #1553708. Any opinions, findings, and conclusions or recommendations expressed in this material are those of the author(s) and do not necessarily reflect the views of the National Science Foundation.

References

Adler, J. (2001). *Teaching mathematics in multilingual classrooms.* Dordrecht: Kluwer. https://doi. org/10.1023/A:1017980828943.

Adler, J., & Ronda, E. (2015). A framework for describing mathematics discourse in instruction and interpreting differences in teaching. *African Journal of Research in Mathematics, Science and Technology Education, 19*(3), 237–254. https://doi.org/10.1080/10288457.2015.1089677.

Barwell, R., Moschkovich, J. N., & Phakeng, M. (2017). Language diversity and mathematics: Second language, bilingual, and multilingual learners. In J. Cai (Ed.), *Compendium for research in mathematics education* (pp. 583–606). National Council of Teachers of Mathematics.

Boaler, J., & Greeno, J. G. (2000). Identity, agency and knowing in mathematics worlds. In J. Boaler (Ed.), *Multiple perspectives on mathematics teaching and learning* (pp. 171–200). Westport, CT: Ablex Publishing.

Callahan, R. M. (2005). Tracking and high school English learners: Limiting opportunity to learn. *American Educational Research Journal, 42*(2), 305–328.

Chapin, S., O'Connor, M., & Anderson, N. (2014). *Classroom discussions in math: A teacher's guide for using talk moves to support Common Core and more.* Math Solutions.

Chval, K. B., Pinnow, R. J., & Thomas, A. (2014). Learning how to focus on language while teaching mathematics to English language learners: A case study of Courtney. *Mathematics Education Research Journal, 27*(1), 103–127.

Design-Based Research Collective. (2003). Design-based research: An emerging paradigm for educational inquiry. *Educational Researcher, 32*(1), 5–8.

Driscoll, M., Nikula, J., & DePiper, J. N. (2016). *Mathematical thinking and communication: Access for English learners.* Heinemann.

Forman, E. (1996). Learning mathematics as participation in classroom practice: Implications of sociocultural theory. In L. Steffe, P. Nesher, P. Cobb, G. A. Goldin, & B. Greer (Eds.), *Theories of mathematics learning* (pp. 115–130). Hillsdale, NJ: Lawrence Erlbaum.

Gándara, P. C., & Contreras, F. (2009). *The Latino education crisis: The consequences of failed social policies.* Cambridge, MA: Harvard University Press.

Gee, J. P. (1996). *Social linguistics and literacies: Ideology in discourses.* London: Taylor & Francis Inc.

Gee, J. P., & Green, J. L. (1998). Discourse analysis, learning, and social practice: A methodological study. *Review of Research in Education, 23,* 119–169.

Gravemeijer, K., & Doorman, M. (1999). Context problems in realistic mathematics education: A calculus course as an example. *Educational Studies in Mathematics, 39*(1/3), 111–129.

Kanno, Y., & Kangas, S. E. N. (2014). "I'm not going to be, like, for the AP": English language learners' limited access to advanced college-preparatory courses in high school. *American Educational Research Journal, 51*(5), 848–878. https://doi.org/10.3102/0002831214544716.

Lave, J., & Wenger, E. (1990). *Situated learning: Legitimate peripheral participation.* Cambridge: Cambridge University Press.

Lobato, J., & Ellis, A. B. (2010). *Developing essential understanding of ratios, proportions, and proportional reasoning for teaching mathematics: Grades 6–8.* National Council of Teachers of Mathematics.

Martiniello, M. (2008). Language and the performance of English-language learners in math word problems. *Harvard Educational Review, 78,* 333–368.

Moschkovich, J. N. (2002). A situated and sociocultural perspective on bilingual mathematics learners. *Mathematical Thinking and Learning, 4*(2–3), 189–212.

Moschkovich, J. N. (2015). Academic literacy in mathematics for English learners. *The Journal of Mathematical Behavior, 40,* 43–62.

Moschkovich, J. N., & Zahner, W. (2018). Using the academic literacy in mathematics framework to uncover multiple aspects of activity during peer mathematical discussions. *ZDM Mathematics Education, 50*(6), 999–1011.

National Research Council. (2001). *Adding it up: Helping children learn mathematics*. Washington, DC: National Academy Press.

Ng, O.-L. (2016). The interplay between language, gestures, dragging and diagrams in bilingual learners' mathematical communications. *Educational Studies in Mathematics, 91*(3), 307–326. https://doi.org/10.1007/s10649-015-9652-9.

O'Connor, C., Michaels, S., & Chapin, S. (2015). "Scaling down" to explore the role of talk in learning: From district intervention to controlled classroom study. In L. B. Resnick, C. S. C. Asterhan, & S. N. Clarke (Eds.), *Socializing intelligence through academic talk and dialogue* (pp. 111–126). Washington, DC: American Educational Research Association.

Prediger, S., Gravemeijer, K., & Confrey, J. (2015). Design research with a focus on learning processes: An overview on achievements and challenges. *ZDM—Mathematics Education, 47*(6), 877–891.

Prediger, S., & Zindel, C. (2017). School academic language demands for understanding functional relationships: A design research project on the role of language in reading and learning. *Eurasia Journal of Mathematics, Science and Technology Education, 13*(7b), 4157–4188.

Roschelle, J., Kaput, J., & Stroup, W. (2000). SimCalc: Accelerating student engagement with the mathematics of change. In M. J. Jacobson & R. B. Kozma (Eds.), *Learning the sciences of the 21st century: Research design and implementing advanced technology learning environments* (pp. 47–75). Hillsdale, NJ: Erlbaum.

Setati, M., & Adler, J. (2000). Between languages and discourses: Language practices in primary multilingual mathematics classrooms in South Africa. *Educational Studies in Mathematics, 43*(3), 243–269. https://doi.org/10.1023/A:1011996002062.

Stump, S. L. (2001). High school precalculus students' understanding of slope as measure. *School Science and Mathematics, 101,* 81–89.

Thompson, P. W. (1994). The development of the concept of speed and its relationship to concepts of rate. In G. Harel & J. Confrey (Eds.), *The development of multiplicative reasoning in the learning of mathematics* (pp. 179–234). Albany, NY: SUNY Press.

Thompson, P. W. (2008). Conceptual analysis of mathematical ideas: Some spadework at the foundation of mathematics education. In O. Figueras, J. L. Cortina, T. Alatorre, T. Rojano, & A. Sepulveda (Eds.), *Proceedings of the Joint Meeting of PME 32 and PME-NA XXX*.

Zahner, W. (2015). The rise and run of a computational understanding of slope in a conceptually focused bilingual algebra class. *Educational Studies in Mathematics, 88*(1), 19–41.

Zahner, W., Calleros, E. D., & Pelaez, K. (2021). Designing learning environments to promote academic literacy in mathematics in multilingual secondary mathematics classrooms. *ZDM Mathematics Education*. https://doi.org/10.1007/s11858-021-01239-0.

Zahner, W., Pelaez, K., & Calleros, E. (2020). Designing mathematics learning environments for multilingual students: Results of a redesign effort in introductory algebra. In A. I. Sacristán & J. C. Cortés (Eds.), *Proceedings of the 42nd annual meeting of the North American chapter of the international group for the psychology of mathematics education*. Mazatlán, MX: PMENA.

Zahner, W., Velazquez, G., Moschkovich, J. N., Vahey, P., & Lara-Meloy, T. (2012). Mathematics teaching practices with technology that support conceptual understanding for Latino/a students. *Journal of Mathematical Behavior, 31,* 431–446.

Zahner, W., Wynn, L., & Ulloa, S. (2018). Designing and redesigning a lesson for equity and access in a linguistically diverse high school classroom. In D. White, A. Fernandes, & M. Civil (Eds.), *Access and equity promoting high quality mathematics in grades 9–12* (pp. 107–124). Reston, VA: National Council of Teachers of Mathematics.

Zahner, W. C. (2012). ELLs and group work: It can be done well. *Mathematics Teaching in the Middle School, 18*(3), 156–164.

Zwiers, J., Dieckmann, J., Rutherford-Quach, S., Daro, V., Skarin, R., Weiss, S., & Malamut, J. (2017). *Principles for the design of mathematics curricula: Promoting language and content development*. Stanford University UL/Scale. http://ell.stanford.edu/content/mathematics-resources-additional-resources.

William Zahner , Ph.D., is an Associate Professor in the Department of Mathematics and Statistics at San Diego State University. Zahner teaches mathematics courses for prospective secondary teachers and is the principal investigator of research projects focused on language diversity and mathematics. Prior to his work as a faculty member, Zahner was a high school mathematics teacher in Chuuk, Federated States of Micronesia and San Jose, California.

Kevin Pelaez , M.S., is a graduate student in the Mathematics and Science Education joint doctoral program at San Diego State University and University of California San Diego. Prior to pursuing a Ph.D., Kevin taught mathematics at the high school and college level. His research interests include teacher education, and equity and justice in education.

Ernesto Daniel Calleros , M.S., is a Ph.D. student in the Mathematics and Science Education joint doctoral program at San Diego State University and University of California, San Diego. Prior to the Ph.D. program, Ernesto taught college-level mathematics to both undergraduates and high school students. His research interests include the teaching and learning of university mathematics and making mathematics accessible to all students, especially multilingual learners.